T0418807

BALANCE DYSFUNCTION IN PARKINSON'S DISEASE

BALANCE DYSFUNCTION IN PARKINSON'S DISEASE

Basic Mechanisms to Clinical Management

MARTINA MANCINI

Assistant Professor of Neurology,
Oregon Health & Science University, Portland, Oregon

JOHN G. NUTT

Professor of Neurology, Oregon Health & Science University, Portland, Oregon

FAY B. HORAK

Professor of Neurology, Biomedical Engineering and Behavioral Neuroscience,
Oregon Health & Science University, Portland, Oregon

Academic Press is an imprint of Elsevier
125 London Wall, London EC2Y 5AS, United Kingdom
525 B Street, Suite 1650, San Diego, CA 92101, United States
50 Hampshire Street, 5th Floor, Cambridge, MA 02139, United States
The Boulevard, Langford Lane, Kidlington, Oxford OX5 1GB, United Kingdom

Library of Congress Cataloging-in-Publication Data
A catalog record for this book is available from the Library of Congress

British Library Cataloguing-in-Publication Data
A catalogue record for this book is available from the British Library

ISBN: 978-0-12-813874-8

For information on all Academic Press publications visit our website at https://www.elsevier.com/books-and-journals

Publisher: Nikki Levy
Acquisition Editor: Melanie Tucker
Editorial Project Manager: Kristi Anderson
Production Project Manager: Bharatwaj Varatharajan
Cover Designer: Miles Hitchen

Typeset by TNQ Technologies

We wish to dedicate this book to all the patients who participated in research studies and have taught us so much about Parkinson's disease.

Contents

7. How and why is turning affected by Parkinson disease?

8. Is freezing of gait a balance disorder?

9. How should the clinician approach imbalance in PD?

10. Future perspectives on balance disorders in PD

About the authors

Martina Mancini, PhD, is currently Assistant Professor of Neurology and Co-director of the Balance Disorders Laboratory at Oregon Health & Science University. Dr. Mancini is a bioengineer focusing on the use of technologies to characterize and treat mobility impairments, such as freezing of gait, in people with Parkinson's disease. She is investigating the neural correlates of mobility changes with technology-based approach for rehabilitation. Dr. Mancini received her BS, MS, and PhD in Bioengineering at the Alma Mater Studiorum, University of Bologna, and her post-doctoral fellowship in Neuroscience at OHSU. Dr. Mancini has authored over 70 peer-reviewed papers.

John G. Nutt, MD, is Co-founder and Director Emeritus of the OHSU Parkinson Center and Movement Disorders Program. Dr. Nutt has gained international recognition for his innovative research in movement disorders and he is recognized for his work on the pharmacokinetics of levodopa which has provided significant clinical and scientific insight on this important therapy for PD. His background in pharmacology has made him a world leader in testing many novel therapeutics for Parkinson's disease as well as new neuroprotective and neurorestorative therapies. He has also become a world leader and international expert in understanding the gait and balance problems of PD and ways to better manage these problems. Dr. Nutt received his MD degree at the Baylor College of Medicine in Houston; he completed his residency in Neurology at the University of Washington in Seattle, and his clinical fellowship at the Addiction Research Institute in Lexington, Kentucky. Dr. Nutt has authored over 350 peer-reviewed papers with international collaborators.

Fay B. Horak, PT, PhD, is a Professor of Neurology, Founder and Co-director of the Balance Disorders Laboratory at Oregon Health & Science University. Dr. Horak is a physical therapist and neuroscientist who is internationally known for her research on the physiology of balance disorders in Parkinson's disease and their rehabilitation. She is an internationally recognized innovator, mentor, and leader in the field of balance disorders characterization. Dr. Horak received a BS degree in physical therapy from the University of Wisconsin, an MS in neurophysiology from the University of Minnesota, and a PhD in Neuroscience from the University of Washington in Seattle. Dr. Horak has authored over 300 peer-reviewed papers, copyrights for balance assessment tools, and patents related to technologies using body-worn sensors to quantify and rehabilitate balance and gait.

Acknowledgments

In bringing together this book we thank the collaborators, fellows, students, post-docs, research assistants from all over the world who have contributed to the cited studies. In addition we are very grateful for the support from the members of the Parkinson Center of Oregon and the Balance Disorders Laboratory at the Oregon Health & Science University.

Introduction

This is the first book to comprehensively summarize the research on how Parkinson disease (PD) affects control of balance and indicates how this research can inform and improve clinical care of patients with PD. We present a framework for understanding balance disorders in PD to help clinicians differentiate types of balance disorders and causes of fall risk, as well as focus therapy for specific balance deficits. In addition, PD affects many different aspects of balance control, so it is a great model for basic scientists to better understand normal control of balance by the brain. Thus, this book is aimed at both clinicians and basic scientists who are working to improve our understanding and treatment of balance disorders associated with PD. It may also be of use to people with PD and their caregivers seeking a deeper understanding of why balance control is such a big problem and what to do about it.

We summarize the results of laboratory studies on four different balance domains affected by PD: (1) postural sway during quiet stance, (2) automatic postural responses to external perturbations, (3) anticipatory postural adjustments in preparation for voluntary movements, (4) dynamic balance during walking. We will use these balance control domains to describe control of balance during turning while walking as well as freezing of gait (FoG). Each chapter also describes how current treatment approaches (medication, deep brain stimulation, and exercise/rehabilitation) have been shown to affect each balance domain.

Over 50,000 Americans are diagnosed with PD each year, with a prevalence of about 1% of the population over the age of 60 and increasing prevalence with age. The major motor disturbances in PD are postural instability, bradykinesia (i.e., slowed movement), rigidity, and rest tremor. However, postural instability largely defines progression of the disease with the "pull test" in the Hoehn and Yahr scale. Quality of life is very strongly related to postural instability and gait in the PDQ-39 Quality of life scale.[1] However, PD does not affect everyone the same way, with different effects on each balance domain for each individual. Although early symptoms of PD emerge gradually and may be subtle, laboratory studies show that some aspects of balance control are affected very early, prior to awareness of the patient or their clinicians to abnormalities of balance dysfunction. In fact, quantitative measures of balance control may provide a powerful biomarker for tracking disease progression and sensitivity of the disease to treatment.

Balance impairments are the most important and modifiable contributor to falls and fall injuries. People with PD are twice as likely to fall as people with other neurological conditions.[2] In addition, around 70% of people with PD who fall do so recurrently and many fall very frequently.[3] The consequences of these falls are significant; they often result in injury, reduce activity levels, reduce quality of life, and increase caregiver stress. Given that the prevalence of PD in developed countries is expected to double from 2005 to 2030,[4] PD-related falls can be expected to have a major impact on healthcare systems in the coming decades. Therefore, it is critical to understand how current and future treatments may improve or worsen balance in PD.

The past 50 years of laboratory study on balance in PD demonstrate that postural control is a complex motor skill derived from the interaction of multiple sensorimotor processes. Balance control was once assumed to consist of a set of reflexes that triggered equilibrium responses based on visual, vestibular, or somatosensory triggers.[5] Likewise, it was assumed that one, or a few, "balance centers" in the central nervous system (CNS) were responsible for the control of balance. This simple view of a balance system is quite limiting and can partially account for our limited abilities to assess risks of falling accurately, to improve balance, and to reduce falls in PD. Sometimes clinicians confuse gait disorders with balance disorders because balance control is an important prerequisite for safe, functional mobility. However, gait (locomotion) and balance are not equivalent, as they are different physical functions requiring different control systems and therapeutic approaches.

We believe that qualitative, global measures of balance are insufficient to identify specific constraints on the balance control to customize intervention for each individual with PD. A better understanding of how the brain controls balance and what goes wrong with balance control in people with PD is the first step in improving quality of life for people with parkinsonism, as well as elderly people who may not have optimal use of their dopaminergic basal ganglia system.

References

1 Schrag A, Jahanshahi M, Quinn N. What contributes to quality of life in patients with Parkinson's disease? *Journal of Neurology, Neurosurgery & Psychiatry* 2000;**69**(3):308–12.

2 Stolze H, Klebe S, Zechlin C, Baecker C, Friege L, Deuschl G. Falls in frequent neurological diseases–prevalence, risk factors and aetiology. *Journal of Neurology* 2004;**251**(1):79–84.

3 Allen NE, Schwarzel AK, Canning CG. Recurrent falls in Parkinson's disease: a systematic review. *Parkinson's Disease* 2013;**2013**:906274.
4 Dorsey ER, Constantinescu R, Thompson JP, et al. Projected number of people with Parkinson disease in the most populous nations, 2005 through 2030. *Neurology* 2007;**68**(5):384–6.
5 Horak FB. Postural orientation and equilibrium: what do we need to know about neural control of balance to prevent falls? *Age and Ageing* 2006;**35**(Suppl. 2):ii7–11.

CHAPTER

1

How is balance controlled by the nervous system?

A. What is balance control?

The balance (posture control) system has two main goals: (1) postural equilibrium (stability) and postural orientation (alignment and sensory orientation). Effective control of both equilibrium and orientation depends upon properties of the balance control system including the musculoskeletal system, antigravity tone, internal model, cognitive control, coupling voluntary and postural goals, and ability to improve control via motor learning. Comprehensive assessment of balance requires examining ability to accomplish a variety of balance tasks/domains that depend upon achieving the goals of equilibrium and orientation: (1) standing, (2) transitioning between postures or between posture and movement, (3) reacting to external perturbations, and (4) walking (straight or turning). This book is organized around these four postural domains to provide clinicians and researchers a framework to consider when assessing balance control in patients with Parkinson disease (PD).

Postural equilibrium (also called balance) involves control by the nervous system to resist forces acting on the body that attempt to alter desired body position (postural alignment).[1] Whereas inanimate objects, like a stacked rock tower can maintain static equilibrium when the center of mass of each segment is lined up over its base of support, humans continuously and actively control dynamic equilibrium by moving either, or both, the body center of mass (CoM) or the base of support. The CoM is a point that represents the average position of the body's total mass. In humans, the CoM is approximately 2 cm in front of the second lumbar spinal segment while standing, but can be in front of the body when flexed at the hips. Because the body is always in motion, even when

Balance Dysfunction in Parkinson's Disease
https://doi.org/10.1016/B978-0-12-813874-8.00001-5

attempting to stand still, the CoM continually moves about with respect to the base of foot support, and has been called postural sway. To control the moving CoM, we push down onto the support surface with our feet to move the body center of pressure (CoP) out beyond the CoM to corral it within the base of foot support. If the CoP is ineffective in quickly controlling the CoM, people take a step or reach out to a stable object to recover equilibrium. Postural instability is determined by how fast the body CoM moves toward the boundary of its base of support and how close the downward projection of the body's CoM is to the support boundary. Thus, holding onto a stable walker can improve control of postural equilibrium by increasing the base of support, allowing the arms to provide forces to control motion of the CoM, and by supplementing somatosensory information from the arms for control of balance.

Postural orientation involves arranging body parts with respect to the environment (gravity, visual and support surface) to accomplish tasks efficiently, interpret sensory inputs, and anticipate balance disturbances. In humans, upright orientation of the trunk with respect to gravity minimizes the forces and energy required to control the body's CoM over its base of support. However, for some tasks the position of a body part in space needs to be stabilized, whereas for other tasks, one body part needs to be stabilized with respect to another. For example, when walking while carrying a glass full of water, it is important to stabilize the hand with respect to gravity to prevent spillage. In contrast, when walking while reading a book, the hand must be stabilized with respect to the head and eyes. Subjects may adopt a particular postural orientation to optimize the accuracy of sensory signals regarding body motion. For example, in activities such as windsurfing or skiing, in which the support surface is unstable, information about earth vertical is derived primarily from vestibular and visual inputs. A person often aligns his head with respect to gravitational vertical in this situation because the perception of vertical is most accurate when the head is upright and stable.[2,3] Anticipatory alterations of habitual body orientation can minimize the effect of an anticipated postural perturbation. For example, people often lean in the direction of an anticipated external force, or flex their knees, widen their stance, and extend their arms when anticipating that stability will be compromised. Curiously, people with PD may not flexibly modify their postural orientation faced with such anticipated perturbations.[4]

Postural equilibrium and postural orientation are different but interdependent. For example, a patient with camptocormia, or involuntary flexion of the hips while standing, can have excellent control of their body center of mass (equilibrium) for both self-initiated and external perturbations (Fig. 1.1A). In contrast, another patient can have excellent postural orientation, both in body alignment and in multisensory orientation to the world, but find themselves falling often due to poor control of

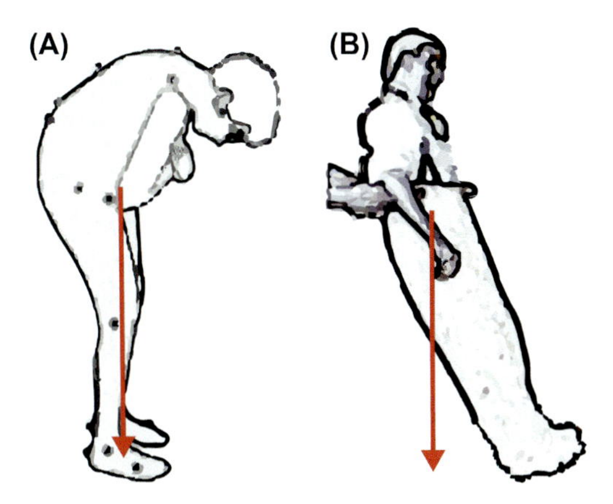

FIGURE 1.1 (A) Poor postural orientation with good control of equilibrium (CoM is over base of foot support) in a patient with camptocormia and (B) Poor equilibrium control with good postural orientation in a patient with rigidity who is falling (projection of CoM to the ground is behind the feet). *(A) Adapted from St George RJ, Gurfinkel VS, Kraakevik J, Nutt JG, Horak FB. Case studies in neuroscience: a dissociation of balance and posture demonstrated by camptocormia.* Journal of Neurophysiology *2017. https://doi.org/10.1152/jn.00582.2017.*

postural equilibrium for self-initiated and/or external perturbations[5] (Fig. 1.1B, Purdue Martin, 1967, book out of print). Although postural equilibrium and orientation can be altered independently, they are also interdependent. For example, studies have shown that a flexed postural orientation of the legs and trunk compromises the ability to recover equilibrium in response to perturbations (Chapter 4).

B. What are the critical properties of balance control?

Underlying effective balance is a control system with specific properties that provides flexibility, resilience, and efficiency to balance control. The properties include the musculoskeletal system, antigravity muscle tone, internal model of the body and world, coupling between posture and movement, and cognitive control of balance. All of these properties can be impaired in people with PD, resulting in difficulty controlling postural equilibrium and orientation across all the tasks they attempt.

Property 1. Musculoskeletal system: Neural control is exerted via a musculoskeletal system than provides some passive, mechanical stability and orientation. For example, standing can be accomplished with only tonic activation of soleus muscles in the calves because ligaments and other elastic structures around the joints and alignment of body segment center of mass over each segment can allow for static equilibrium. However, abnormal musculoskeletal system due to reduced joint range, reduced muscle strength, unusual segmental body alignment, etc., usually results in inefficient, less effective balance control. Thus, the typical flexed posture of a person with PD puts them at a disadvantage for efficient control of balance (see Chapter 3).

Property 2. Antigravity muscle tone: Every posture and every movement made in a gravitational environment requires antigravitational support to prevent the joints from collapsing. However, postural muscle tone is not a static, steady state, but rather a flexible support that varies depending on the desired actions, environmental context, and biomechanical requirements. Like the waves of an ocean supporting a floating vessel, postural tone provides a constantly changing background stiffness around which movement and postural goals can be accomplished efficiently and effectively. For example, while standing upright against a strong wind, background tonic activity of a large set of muscles resists the forces but must change abruptly when walking or other voluntary movement is initiated. Thus, the inflexible, often increased, flexor tone associated with rigidity of PD may provide good antigravity support, but endangers postural stability and orientation as observed by falls and inefficient, painful postural alignment (see Chapters 3 and 7).

Property 3. Internal model of body and world: The nervous system forms an internal representation, or 'gestalt', of the physical properties of the body and world that is used to issue appropriate, feedforward commands to muscles.[6] This internal model includes estimations of body's physical dimensions, physics of movement in each particular environment, and the relationship between the body and the environment that is the basis for control of every posture and movement. For example, this internal model provides the basis for representation of the "limits of stability" or how far the body center of mass can move without changing the base of support. The internal model also allows construction of representations of postural (proprioceptive) and visual vertical as goals for postural orientation/alignment. Thus, a patient with PD who leans to one side with Pisa syndrome may have an abnormal internal model of the body or world (see Chapter 3).

Property 4. Coupling between balance and movement: Although neural control of posture is necessary to support effective and efficient movements, such as walking and reaching, neural control of balance and of movement are not the same. For example, there are neurons in the reticular formation that are active in association with postural adjustments prior to taking a step and different neurons that are active in association with the step itself.[7] Thus, a complex spatial-temporal coupling must occur between commands for balance control and commands for movement control. This coupling is complex because any muscle can be used either or both for control of balance and/or movement. Coupling is also complex because the critical timing between activation of muscles to control balance and movement can be within milliseconds to provide effective and efficient control. Thus, a patient with freezing of gait who has difficulty coupling anticipatory postural adjustments prior to a voluntary step may be revealing abnormalities in coupling between the balance control and gait movement control systems (see Chapters 4 and 8).

Property 5. Cognitive control of balance: A 100 years ago, balance was thought to be controlled via spinal and/or brainstem reflexes, like the righting and stretch reflexes.[8] However, now we understand that control of balance is a complex, sensorimotor skill that is learned during development and can be improved with practice. To learn to balance in a new condition or to accomplish a novel task requires cortical control via reward-based learning via the basal ganglia, and error-based learning via the cerebellum. However, as the balancing task becomes more automatic, cortical control of balance is minimized until an unexpected change in circumstances or expected change in intended movement or posture is initiated. Thus, healthy people have little difficulty accomplishing tasks such as walking, standing, transitioning between postures, and responding to perturbations while simultaneously accomplishing a cognitive task because their well-practiced balancing tasks are under automatic control. In contrast, a person with PD may find it difficult to walk, stand, transition, or respond while talking (see Chapters 6 and 8).

Fig. 1.2 summarizes these properties of the balance control system with a neural control model. Given a desired balance state (e.g., particular

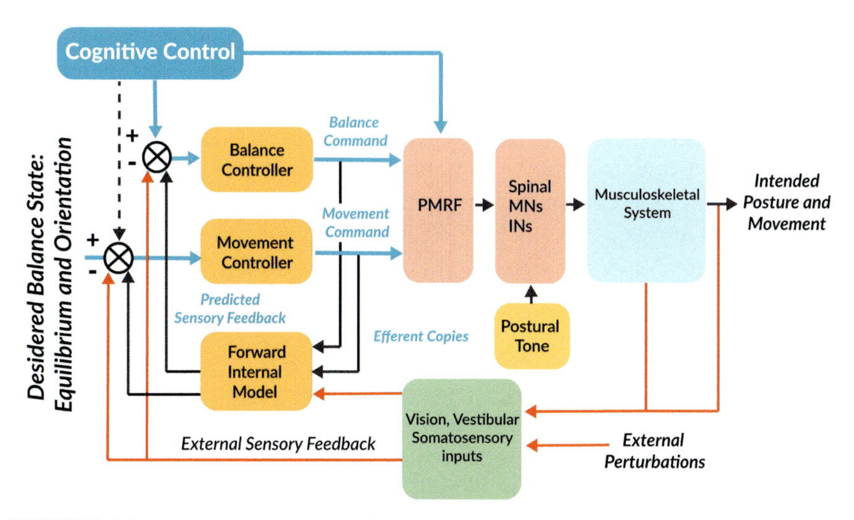

FIGURE 1.2 Neural control model for posture and movement. *Adapted from Mackinnon C. Sensorimotor anatomy of gait, balance, and falls. In:* Handbook of clinical neurology. *Elsevier; 2018.*

postural equilibrium and postural orientation) and the particular cognitive command, the balance controller and movement (such as locomotion or reaching) controllers are coupled via an internal model of the body and the world. Output of the combined balance/movement command to the brainstem muscle synergy centers (Pontomedullary Reticular Formation, PMRF) and to spinal motoneurons (MNs) and interneurons (INs) are then implemented over background postural tone and executed via the musculoskeletal system. Sensory feedback from vision, vestibular, and somatosensory systems, as well as expected sensory feedback, are used to maintain steady state balance, balance during voluntary movement, as well as respond to perturbations. PD affects every one of these properties of the balance control system as will be apparent in the chapters that follow.

C. What are the main balance domains to assess?

Balance Control involves controlling equilibrium and orientation under a variety of tasks or domains. Balance is not a reflex or single brain function, but consists of a set of complex set of sensorimotor skills that are involved in most voluntary and involuntary movements and all movements of the body through space. Clinicians and researchers who need to comprehensively characterize balance control in patients with PD will need to examine behaviors across several domains using different tasks as each of these domains can be affected without necessarily affecting the other domains. This section will summarize the studies in healthy human subjects characterizing four domains of postural control: (1) postural sway during standing, (2) postural responses to external perturbations, (3) anticipatory postural adjustments prior to voluntary movement and postural transitions, and (4) dynamic postural control during locomotion. Fig. 1.3 shows how these four domains result in what is sometimes called "Balance," supported by critical properties of the postural control system, as discussed above.

Standing balance

Standing without external support is an active process involving closed-loop sensory feedback.[10] Sensory information from the somatosensory (proprioceptive and cutaneous), visual, and vestibular systems are integrated to form an internal representation of body orientation in space (with respect to surface, visual, and gravitation/inertial cues). The balance control system changes the relative sensitivity or weighting of each sensory input to accommodate changes in the environment and movement goals. Subjects on a firm, stable surface tend to rely primarily

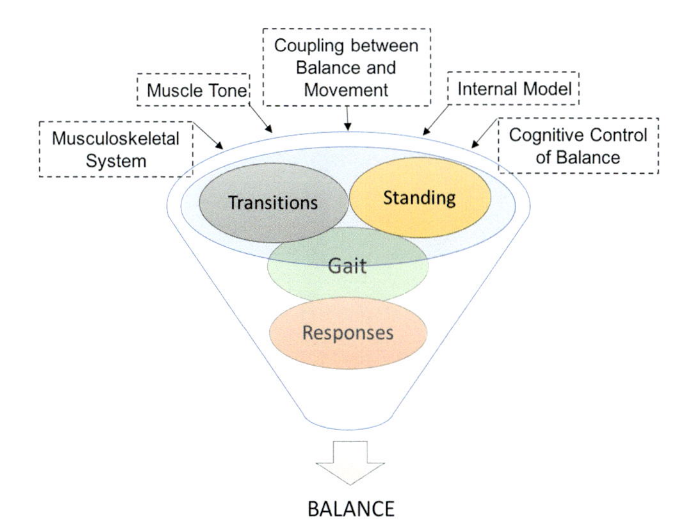

FIGURE 1.3 Balance domains and related properties. Four main domains of posture control for control of standing, postural transitions, gait and postural responses result in "balance." All of these domains depend upon having an adequate musculoskeletal system, muscle tone, coupling between balance and movement, an internal model of the body, and environment and cognitive control.

(70%) on somatosensory information from the support surface for postural orientation. However, when the support surface is unstable, subjects depend more on vestibular and visual information than somatosensory information from the moving surface, which no longer provides reliable information about body CoM motion.[10] In addition, light touch with a fingertip is even more effective than vision/vestibular in maintaining postural orientation and equilibrium when standing on an unstable surface.[11,12] Vestibular information is particularly critical when both visual and somatosensory information is ambiguous or absent, such as when skiing downhill or walking below deck on a ship.

Postural stability while standing unsupported can be measured with a force plate as displacement of center or pressure or with an accelerometer on the pelvis as change in body acceleration in the forward-backward and lateral directions. Generally, the larger, faster, and higher the frequency of postural sway, the worse is the postural stability. Postural sway can be measured across several visual and/or surface conditions to estimate relative reliance upon visual and/or surface inputs and ability to use vestibular information when both vision and surface information is compromised (Fig. 1.4A). For example, comparing sway with eyes closed versus with eyes open provides an estimate of visual dependence for standing postural stability.

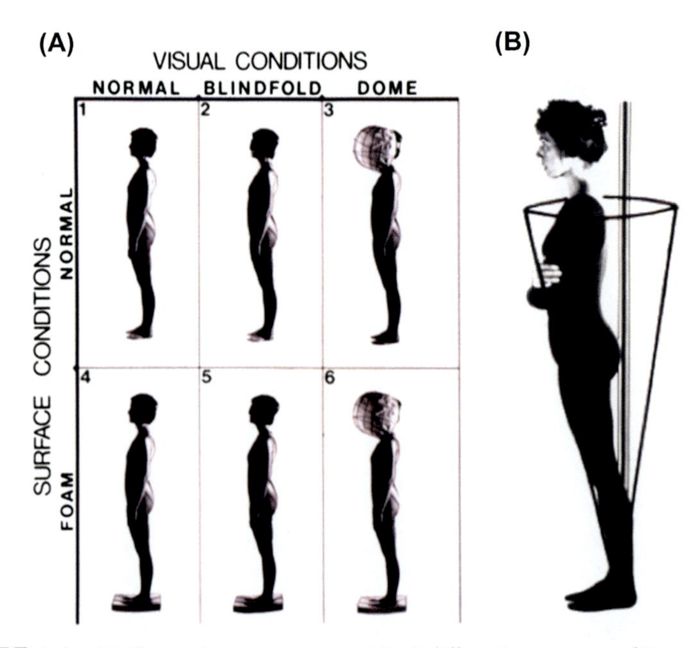

FIGURE 1.4 (A) Postural sway assessment in 6 different sensory conditions, ordered from easiest to most difficult. Conditions 1–3 are on a firm surface with eyes open, eyes closed, and with a head-fixed visual surround. Conditions 4–6 are standing on a compliant foam surface in the same 3 visual conditions. (B) Cone of postural stability in a healthy adult based on their maximum body tilt in each direction. *Adapted from Horak.* Physical Therapy *1987.*

When subjects are asked to tilt their body to the limits of their stability while standing, this forms a cone shape of stability limits (Fig. 1.4B). Limits of stability are biomechanically limited to approximately 8 degrees forward and 4 degrees backward, with lateral limits dependent on the width of foot support. However, limits of stability are dependent upon musculoskeletal constraints, such as strength and joint range, as well as upon the internal model of body and world. For example, subjects with fear of falling, but normal musculoskeletal system, may reduce their voluntary limits of stability to keep their body center of gravity well within their limits of foot support.

Postural responses (APRs)

To return the body back to equilibrium and desired orientation in response to external disturbances (perturbations), the nervous system relies upon automatic postural responses. If the perturbation is small, humans use an "ankle strategy" in which the body rotates as an inverted pendulum with forces and movements primarily about the ankle joints (Fig. 1.5). Larger perturbations, in which the body CoM cannot return to equilibrium by exerting torque about the ankles, results in either a "hip strategy" or "stepping strategy." A hip strategy uses rapid flexion/extension about the hips to more quickly move the body CoM than an ankle strategy. The disadvantage of the hip strategy is that postural orientation in space is compromised, with the head moving in the opposite direction as the body CoM correction. A postural stepping strategy has a longer latency than either an ankle or hip strategy as generally a subject first attempts one of the feet-in-place strategies prior

FIGURE 1.5 Postural strategy representation. In response to forward disequilibrium. (A) Ankle strategy, (B) Hip strategy, and (C) Stepping strategy. In response to lateral disequilibrium, (D) Trunk vertical with hip abduction/adduction strategy, (E) Lateral trunk strategy, and (F) Stepping strategy. *Adapted from Horak and Shumway-Cook, APTA Posture Symposium 1989.*

to eliciting a step. Stepping is also slower (latency to first foot off the ground over 250 ms after a perturbation) because it may require anticipatory weight shifting to a stance leg in order to provide time for the stepping leg to lift off the ground.[13] Stepping strategies are often accompanied by rapid reaching strategies in which an arm will reach for a stable surface to recover equilibrium (arm muscle latencies about 160 ms).

Postural responses include short, medium, and long latency components of muscle activation after a perturbation. Short-latency stretch and force reflexes mediated by spinal reflexes from muscle spindles and Golgi tendon organs activate lower leg muscles at 50 ms but don't generate enough torque to control body CoM. Medium latency postural responses at 75-100 ms likely involve the midbrain/brainstem and long-latency response greater than 100 ms have time enough to involve the cerebral cortex[14] (Fig. 1.6). Thus, automatic

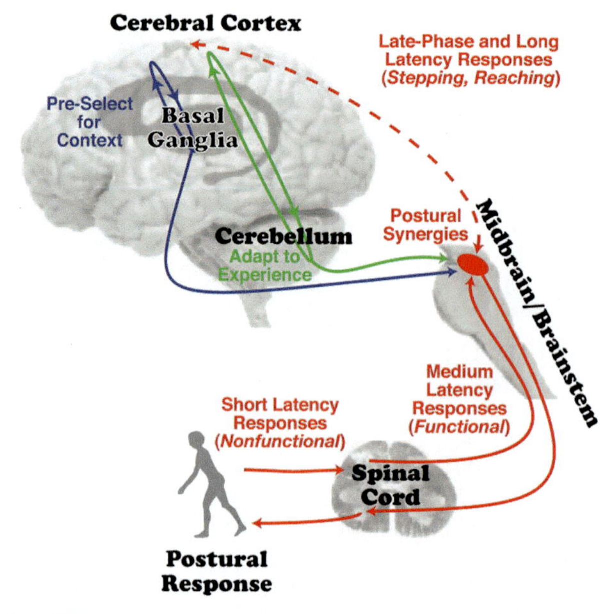

FIGURE 1.6 Proposed neural pathways model involved in cortical control of short, medium, and long latency automatic postural responses. The shortest latency, stretch reflexes are via the spinal cord and are too weak to be functional for postural control. The medium latency responses are via the brainstem and are responsible for the functional, 100 ms latency feet-in-place postural responses. The long latency postural responses are via the sensorimotor cortex and include stepping and reaching responses. The basal ganglia preselect the appropriate postural strategies for the current environmental context and the cerebellum adapts postural responses, so they are optimally based on prior experience. *From Jacobs JV, Horak FB. Cortical control of postural responses.* Journal of Neural Transmission *2007;**114**(10): 1339–48.*

responses to postural perturbations occur more quickly than the fastest voluntary reaction times in the leg; 100 ms latency of gastrocnemius in response to a perturbation versus 180 ms reaction time response of the same muscles to an external sensory cue. Nevertheless, the functionally relevant postural responses to external perturbations have significantly longer latencies than short latency stretch reflexes. Long latencies through the cerebral cortex mean that, although automatic, postural responses are flexible, with great potential for modification by the cortex.

Evidence suggests that the cortex is also involved in adapting postural responses with alterations in cognitive state, initial sensory-motor conditions, prior experience, and prior warning of a perturbation, all representing changes in "central set." Studies suggest that the cerebellar-cortical loop is responsible for adapting postural responses based on prior experience and the basal ganglia (BG)-cortical loop is responsible for preselecting and optimizing postural responses based on current context, based on stored patterns selected by the BG. Thus, the cerebral cortex likely influences longer latency postural responses both directly via corticospinal loops and shorter latency postural responses indirectly via communication with the brainstem centers that harbor the synergies for postural responses, thereby providing both speed and flexibility for preselecting and modifying environmentally appropriate responses to a loss of balance.

Anticipatory postural adjustments (APAs)

Prior to all rapid limb and body motions when unsupported externally, the nervous system executes postural adjustments that anticipate likely postural perturbations that will be induced by the action. One of the first examples of APAs in the literature was activation of leg and back muscles prior to rapid arm elevation when standing.[15] Other examples are the forward and lateral weight shift prior to step initiation, forward weight shift prior to rising onto toes, lateral weight shift prior to one-foot standing, and forward trunk motion prior to rising from a chair.[16] Fig. 1.7 illustrates an APA prior to forward step initiation as a lateral and backward center of pressure displacement toward the stepping leg, accompanied by a lateral and forward CoM displacement toward the stance leg.[16]

Each APA is very specific to the particular perturbation associated with the action. For example, whereas posterior leg muscles are activated prior to a rapid arm elevation, anterior leg muscles are activated prior to a rapid arm descent. When postural stability will not be compromised by the action, the APAs will be absent, such as in rapid arm elevation when

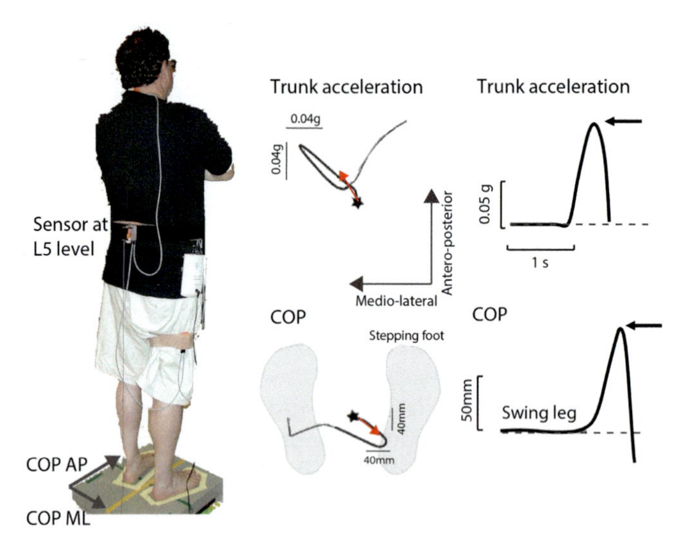

FIGURE 1.7 Anticipatory postural adjustment prior to step initiation as measured from trunk acceleration (top) and from the forceplate center of pressure (bottom). The red arrow shows the initial direction of displacement of the same APA as shown on the right as lateral trunk acceleration and CoP displacement over time. *Adapted from Mancini M, Zampieri C, Carlson-Kuhta P, Chiari L, Horak FB. Anticipatory postural adjustments prior to step initiation are hypometric in untreated Parkinson's disease: an accelerometer-based approach.* European Journal of Neurology *2009;**16**(9):1028–34.*

leaning against a stable support.[17] Since the voluntary movement is delayed until the APA is expressed, elimination of APAs via external stabilization results in faster limb reaction times.[17]

Balance during gait

Human walking represents an extremely complex motor control task that requires challenging control of dynamic balance while attempting forward progression. One balance challenge is that humans are bipeds who usually spend 80% of their walking time on one limb. Thus, only two 10% double support periods are available for restabilizing equilibrium and this time decreases as cadence increases until we break into a run, when it becomes zero double support time and is replaced by a free flight period. Of course, subjects at high fall risk tend to compensate by altering their walking strategy by increasing the duration of their double support time.[18] Another balance challenge while walking is the distribution of body mass, such that two-third of the body mass in the head, arms, and trunk (*H.A.T.*) is located two-third of the body height above the ground. Such an "inverted pendulum" is inherently destabilizing, especially when we consider the body's forward momentum. A third balance challenge is that the projection of the body's CoM to the ground (center of gravity) does not pass within the foot; rather, it passes forward of the foot and along the medial border of the foot.[19] The foot, especially in the medial-lateral direction, provides a very small base of support. Compounded to this balance task is the requirement to achieve a safe forward trajectory of the swing limb with minimum toe clearance of less than 2 cm and a safe (gentle) foot landing.[20,21] To minimize disturbance of visual and vestibular sensory information while walking, inertial disturbances from the feet in contact with the ground need to be suppressed by the time they reach the head.

Thus, body balance during walking is ensured by a continual state of dynamic imbalance toward the center line of the plane of progression with the body CoM passing medial to, and in front of, the supporting foot. Control of the medial acceleration of the CoM is generated by a gravitational moment about the supporting foot, whose magnitude is established at initial contact by the lateral placement of the new supporting foot relative to the horizontal location of the CoM. Balance of the trunk and swing leg about the supporting hip is maintained by active hip abduction and lateral trunk moment, which counters a large destabilizing gravitational moment. However, lateral posture control is also assisted by the medial-lateral placement of the foot with some assistance by the subtalar joint. Interactions between the supporting foot and hip musculature permit variability in strategies used to maintain balance.[22]

When the trajectory of walking needs to change from straight ahead to turning, an even more complex multisensory integration and multijoint coordination strategy is required. Turning generally starts with reorienting the eyes with a rapid eye movement (saccade) that focuses attention to the progression direction. Eye movements are followed by head, then trunk, then pelvis, and leg rotations in a top-down coordination. Turning the head results in modifying the reference frame of visual, vestibular, and neck proprioceptive sensory inputs for balance control. Lateral weight shifting in anticipation of lifting one foot off the ground becomes dangerous as the pivoting support foot on the inside of the turn needs to support the center of gravity accelerating toward the lateral limits of the foot. Thus, falls during turns generally result in landing on the hip in the direction of the turn, with hip fractures common.[23]

D. What parts of the brain are involved in balance control?

It is not known which specific brain networks control balance, but many parts of the brain are involved. Although early studies suggested that neural control of posture relied on spinal reflexes,[24] recent studies have demonstrated the importance of cortical, subcortical, and brainstem areas.[14] Human and animal studies show that postural control involves most of the brain, including the higher-level, prefrontal, and temporal-parietal areas, as well as sensorimotor and subcortical areas of the thalamus, cerebellum, brainstem, as well as the basal ganglia.[14,25,26] However, most anatomical and brain imaging studies with imagined motor functions involve locomotion, with very few involving postural control.[27]

Fig. 1.8 illustrates the distributed nature of brain control of balance with different networks involved in APA, APR, Sway, dynamic balance in gait, and cognitive control of balance. Although specific networks for each of these types of balance control are unknown, we have placed them around different cortical areas. For example, APAs are known to involve control by both the supplementary motor area (SMA) and by the peduncolopontine nuclei (PPN). Inhibition of SMA/preSMA, but not primary motor cortex (M1), with low frequency, continuous transcranial magnetic stimulation (TMS) impairs APAs and patients with lesions in SMA show absence of APAs.[28] APRs are known to be involved the M1-S1 and prefrontal cortex because sudden perturbations in standing humans results in an EEG signal 50 ms after the perturbations with the postural response loop in ankle muscles at 100 ms.[14] Multisensory integration (visual, auditory, and somatosensory) for control of postural sway in standing balance involves the temporal-parietal junction, or "vestibular cortex" inferior frontal gyrus, insula, cingulate, and SMA as shown in monkeys who have difficulty weighting sensory inputs for postural control.[29]

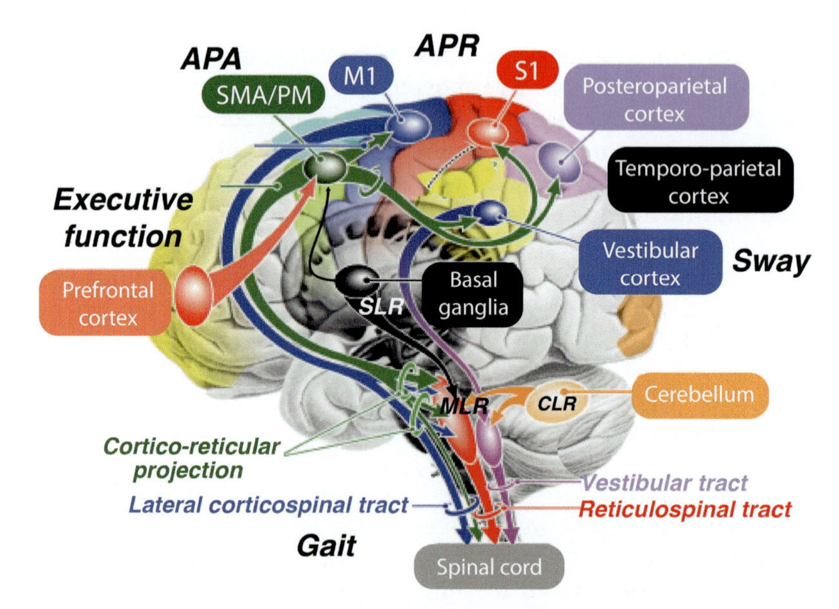

FIGURE 1.8 Brain areas involved in balance control. Many different networks are involved in control of APAs, APRs, postural sway in standing, dynamic balance control during gait, and cognitive control of balance. *Adapted from Takakusaki K. Neurophysiology of gait: From the spinal cord to the frontal lobe, Movement Disorders, 28(11),1483-1491, 2013.*

Dynamic control of balance during gait involves a complex coupling between locomotor pattern generators in the spinal cord and (1) anticipatory postural adjustments for appropriate weight shifts, (2) automatic postural responses to unanticipated perturbations while walking, and (3) sensory integration or orient and balance the head-trunk-arms sections of the body over the moving legs. This integration of locomotor and postural goals likely occurs in the pontomedullary reticular formation and reticulospinal tract, a critical hub for sensorimotor integration that allows the nervous system to appropriately couple voluntary actions with posture and locomotion. This system has been shown to play a prominent role in: (1) anticipatory and reactive postural adjustments; (2) control of locomotor intensity and mode (walking vs. running); and (3) regulation of muscle tone.[7,26,30,31] Modifying gait for turning normally involves planning by the premotor and superior parietal cortex whereas people with PD and freezing of gait seem to preferentially activate visual cortex and inferior frontal regions that have been implicated in the recruitment of a putative stopping network.[32]

The brainstem is critical for many aspects of postural control including control of postural tone, postural sway, muscle synergies for APR, APA coupling with gait initiation and locomotion. Three main areas have been studied in cats and monkeys, the mesancephalic locomotor region (MLR), pontomedullary locomotor region (PLR), and the cerebellar locomotor region (CLR). Although these areas are called "locomotor regions" because that is how they were initially investigated in decerebrate cats, it is clear they have very important postural functions. For example, the MLR consists of the PPN and cuneate nucleus, with the PPN divided into at least three areas with different neurotransmitters and projections. Since the MLR received direct inputs from the globus pallidus and subthalamic nucleus (STN), it is the primary means by which (1) the BG influence control of posture and (2) BG disorders like PD impairs balance control. In fact, a recent study showed that the brainstem may be the most critical area for control of balance because decline in the brainstem structure accounted for age-related decline in postural control irrespective of age, weight, height, static versus dynamic balance conditions, task difficulty, visual condition, ankle joint mobility, and total physical activity.[33] However, brain imaging studies show strong associations between postural performance and many brain regions, encompassing almost the entire brain,[34–38] cerebellum,[39] basal ganglia,[40] and brainstem.[41–43]

The brainstem is the most critical brain area involved in balance control, as it receives input from cortical areas and projects to the spinal cord via the reticulospinal and vestibulospinal pathways for control of balance. Voluntary, goal-directed movements are generally accompanied by automatic processes of postural control, including balance and postural muscle tone. The desire to move or stand still is usually derived from intentionally elicited motor commands from the frontal cortex. However, moving and standing still can also be generated from emotions generated by the limbic-hypothalamus (i.e., fight or flight reactions).

Many cortical areas are involved in postural control. In the last two decades, functional neuroimaging studies in humans have been performed in order to gain a greater understanding of the supraspinal control of balance and walking. Tasks have included both active and mental imagery of stance, walking on a narrow beam, and running.[44] Although the majority of these studies have focused on imaging of walking,[27,45] cortical activation is widely distributed, and greatly task dependent. Imaging studies involving balance tasks have found that frontal areas such as the dorsolateral prefrontal cortex, superior and inferior frontal gyrii, and precentral gyrus are involved in balancing.[35,39,46,47] Also, the parieto-insular vestibular cortex, superior and inferior temporal gyrii, inferior parietal lobe, anterior vermis, precuneus and thalamus are also activated.[39,46,47] Recently, the dorsal and ventral attention network, salience network, and default mode networks have also been shown to be activated for control of postural tasks. For example, EEG studies show activation of the anterior prefrontal cortex, as well as the sensorimotor cortex and SMA in response to external perturbations to standing posture.[48]

Although different brain circuits likely control each postural domain, no single brain area is responsible for any domain of postural control. Studies of APAs prior to step initiation in cats provide a good example of how various parts of the brain work together for postural control.[7] Studies have shown that the pontomedullary reticular formation have neurons active for the APA, other neurons for gait and a third set of neurons activity during both the postural preparation and the stepping. Thus, the brainstem provides coupling of signals from the SMA signaling the size and strategy for the APA with signals from the frontal cortex after context-specific strategy selection by the frontal parietal cortex. The direct outputs of the BG to the SMA and the brainstem, as well as loops to and from all cortical areas put the BG at a critical hub for sizing, adapting, and coupling APAs.

E. What is the role of the basal ganglia in balance control?

The basal ganglia are involved in many aspects of balance control. Since the basal ganglia neither receive direct sensory input nor project directly to the spinal motoneurons, they are primarily involved in the feedforward control of balance and movement. Thus, it is critical for coupling self-initiated movements with balance control. However, disruptions to the internal model also affect postural responses, standing posture, coupling of balance with movement. The basal ganglia exert their influence to posture control via gamma-aminobutyric acid (GABA)-ergic projections to the internal Globus Pallidus (GPi) and Subtantia Nigra pars reticulata (SNr) which in turn have GABAergic projections to the brainstem and to the thalamus, that projects to cortex via glutamatergic signaling. The degree of GABAergic influence from the basal ganglia is regulated by the dopaminergic neurons in the midbrain that begin to degenerate prior to a diagnosis of PD.

The primary roles of the basal ganglia in motor control, and, thus, in balance control are: (1) Energization/Scaling, (2) Automatization, (3) Posture-movement coupling, and (4) Context-dependent adaptation (Fig. 1.9).

1) *Energization/Scaling* (vigor) involves generating appropriate force and rate of change of force for a particular task and condition.[49,50] Each domain of balance control requires appropriate scaling up and down forces appropriate for specific alterations to tasks and conditions. Bradykinesia, or slow force generation, associated with PD results in bradykinetic forces controlling standing, postural transitions, reactions, and walking (Chapters 3–7).
2) *Automatization* involves making habitual tasks more automatic, that is less dependent on cortical attention mechanisms.[51] Early in practice of a new task, frontal and other cortical areas optimize task performance based on error feedback from the cerebellum, as well as reward feedback from the basal ganglia. However, as the tasks, such as walking and standing, become more automatic (motor habits), the basal ganglia is responsible for pushing control from the cortical-BG look to the more automatic, brainstem controllers. This automatization allows for more efficient control of balance so other tasks can be accomplished with the automatic support of balance control. A hallmark of PD control of balance is the excessive need for conscious attention (Chapters 6 and 8).

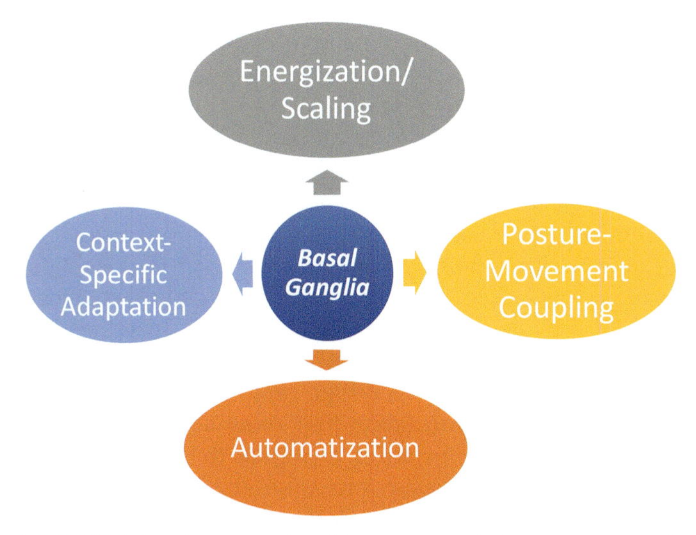

FIGURE 1.9 The main roles of the basal ganglia for posture control.

3) *Posture-Movement Coupling* links the goals of postural stability/orientation with the goals of voluntary/involuntary movement (such as locomotion). This coupling between posture and movement is thought to occur at both brainstem integrative centers such as the reticular formation, as well as at cortical premotor centers such as the SMA.[31] The basal ganglia project to both of these centers and patients with PD have difficultly coupling posture and movement, such as sequencing complex tasks involving postural transitions (Chapter 5).
4) *Context-Dependent Adaptation* of balance control involves modifying muscle synergies and strategies given a particular context. Contexts include environments (i.e., surface, support, visual, etc.), anticipated conditions (i.e., upcoming perturbations and changes in environment, etc.), and intended actions (i.e., desire to remain in place or allow movement). To be efficient and effective, posture control needs to be flexible given changes in context. For example, whereas postural responses to surface displacements start at the ankles when freestanding, they should start at the hands when holding a stable support.[17] Patients with PD tend to be inflexible in their initial choice of postural strategy, based on feedforward control when conditions change, although they are able to modify their strategy trial-by-trial, based on error feedback control.[52]

The basal ganglia receive no direct sensory input and do not project directly to the spinal cord. However, they exert powerful influence over balance control in two ways: via the frontal cortex and via the brainstem. The frontal loop is thought to be involved in less automatic, more voluntary, and learning aspects of balance and movement control, as well as in the coupling between balance and movement. In contrast, the brainstem projection is thought to be involved in more automatic control of (already learned) balance and movement, as well as in control of postural tone and locomotor rhythm.

In the frontal cortex loop, the output nuclei of the BG, theGPi, and the STN project to the ventral lateral and ventral anterior nuclei of the thalamus adjacent to neurons that receive inputs from the cerebellum (Fig. 1.10). These neurons then project to the SMA that in turn projects to the motor cortex. Both SMA and M1 project down to the reticulospinal, vestibulospinal, and corticospinal systems to control posture. In the brainstem control system, the BG projects to the brainstem nucleus, PPN (also called the medullary locomotor region or MLR), particularly the caudal part.[53] This part of the MLR then projects to the medial part of the pontomedullary reticular formation that show activity related to anticipatory and reactive postural control, as well as to locomotion.[7,30,54,55] The medial pontomedullary reticular formation also receives input from the motor regions of the frontal cortex, including the primary motor cortex, premotor cortex, and supplementary motor area.[56–59] This organization

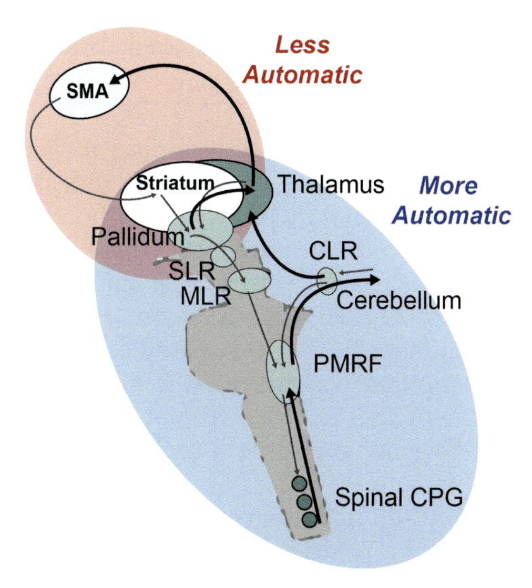

FIGURE 1.10 Supraspinal locomotor centers for balance and gait. The basal ganglia are involved in moving control of balance and gait from "less automatic," that is involving attention from the cortex, to "more automatic," that is, without constant cortical control. *Adapted from Jahn K, Zwergal A. Imaging supraspinal locomotor control in balance disorders.* Restorative Neurology and Neuroscience *2010;***28***(1):105–14.*

provides higher-level control and the ability to couple voluntary motor commands with appropriate postural adjustments.

Although characterization of balance impairments in people with PD can help us understand the role of the basal ganglia in balance control, we must remember that these impairments reflect not only the primary impairment due to the pathophysiology (such as loss of dopamine or particular neural networks), but also reflect adaptive or maladaptive mechanisms by the nervous system to attempt to meet behavioral goals with postural support, given disruption to the system. For example, Fig. 1.11 summarizes how structural changes to the brain result in changes in adaptive or maladaptive brain functional connectivity that expresses itself in altered balance control that then leads to impaired mobility (that is, an integration of gait and balance) that allows for functional, voluntary movements and movement of the body through space.

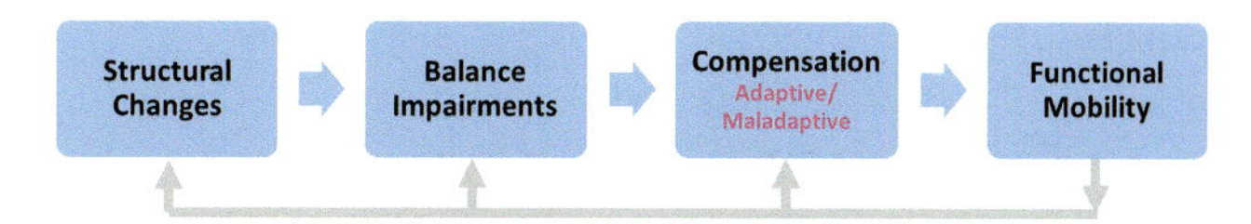

FIGURE 1.11 Model of how degeneration of brain structures due to PD result in balance impairments as well as brain compensatory mechanisms that lead to changes in functional mobility. *Adapted from Papegaaij S, Taube W, Baudry S, Otten E, Hortobagyi T. Aging causes a reorganization of cortical and spinal control of posture.* Frontiers in Aging Neuroscience *2014;***6***:28.*

F. Balance is a complex sensorimotor task that can be improved with practice

Although voluntary and automatic stepping movement are present very early after birth, infants take about a year to learn how to balance well enough to start functional mobility, although falls are still common for years. Each time an individual attempts mobility in a novel set of conditions, such as hiking up a slippery slope, the balance control system adapts to the new context based on both feedforward estimates of what is needed (based on experience), as well as based on feedback control of the error signals between what was intended (efficient mobility) and what actually occurred (inefficient or effective mobility). With repeated practice, balance skills develop, such that a repertoire of potential postural strategies can be automatically called up for each context. Each type of balance domain (standing, transitions, reacting, and walking) has been shown to improve with practice in young, older, and people with neurological disease (albeit less effectively and with more practice required). A good example of improvement of balance control with practice comes from studies showing that rapid, postural stepping responses to external perturbations improve in efficiency with even 1 day of practice and this adaptation becomes learned as there is some retention of the next day. Fig. 1.12 illustrates how a group of 12 elderly subjects decreased their CoM displacement and reduced the number of steps required to recover equilibrium over 25 repeated surface translations that

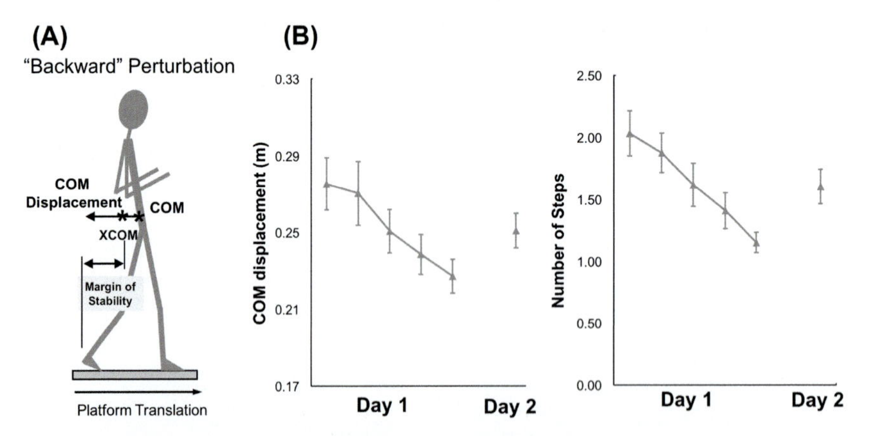

FIGURE 1.12 Learning over 25 trials (mean and SD 5 trials) on 1 day and retention the next day of backward postural stepping responses to surface translations in healthy elderly people.[61] Stick figure defines outcomes such as (A) CoM displacement (movement) in response to perturbations and; (B) Number of steps required to stop falling. The mean and SE of the body CoM movement and number of steps were averaged over 10 trials for five healthy elderly adults and decreased gradually with practice.

caused them to fall backwards.[61] Retention of improvements 24 h later indicated learning had taken place. The following chapters summarize evidence that people with PD can also learn to improve various domains of balance control, albeit not as well as people without PD.

Highlights

- Balance (Posture) control has two main goals: postural equilibrium (stability) and postural orientation.
- Balance control has the following properties: the musculoskeletal system, antigravity postural tone, forward internal model of the body and world, coupling between posture and movement, and cognitive control of balance.
- The main balance domains to assess are: standing balance, postural responses, postural transitions, dynamic balance during gait and turning.
- Control of balance is distributed throughout the brain.
- Balance can be improved with practice.

References

1. Macpherson J, Horak FB. Posture. In: *Principles of neural science*. 5th ed. The McGraw-Hill Companies I; 2013.
2. Dakin CJ, Rosenberg A. Gravity estimation and verticality perception. *Handbook of Clinical Neurology* 2018;**159**:43–59.
3. Wapner S, Werner H. Experiments on sensory-tonic field theory of perception. V. Effect of body status on the kinesthetic perception of verticality. *Journal of Experimental Psychology* 1952;**44**(2):126–31.
4. Shaw JA, Stefanyk LE, Frank JS, Jog MS, Adkin AL. Effects of age and pathology on stance modifications in response to increased postural threat. *Gait and Posture* 2012; **35**(4):658–61.
5. St George RJ, Gurfinkel VS, Kraakevik J, Nutt JG, Horak FB. Case studies in neuroscience: a dissociation of balance and posture demonstrated by camptocormia. *Journal of Neurophysiology* 2017. https://doi.org/10.1152/jn.00582.2017.
6. Ivanenko Y, Gurfinkel VS. Human postural control. *Frontiers in Neuroscience* 2018;**12**:171.
7. Schepens B, Drew T. Independent and convergent signals from the pontomedullary reticular formation contribute to the control of posture and movement during reaching in the cat. *Journal of Neurophysiology* 2004;**92**(4):2217–38.
8. Burke RE. Sir Charles Sherrington's the integrative action of the nervous system: a centenary appreciation. *Brain: A Journal of Neurology* 2007;**130**(Pt 4):887–94.
9. Mackinnon C. Sensorimotor anatomy of gait, balance, and falls. In: *Handbook of clinical neurology*. Elsevier; 2018.
10. Peterka RJ. Sensorimotor integration in human postural control. *Journal of Neurophysiology* 2002;**88**(3):1097–118.
11. Dickstein R, Peterka RJ, Horak FB. Effects of light fingertip touch on postural responses in subjects with diabetic neuropathy. *Journal of Neurology Neurosurgery and Psychiatry* 2003;**74**(5):620–6.

12. Lackner JR, Rabin E, DiZio P. Stabilization of posture by precision touch of the index finger with rigid and flexible filaments. *Experimental Brain Research* 2001;**139**(4):454–64.
13. Peterson DS, Dijkstra BW, Horak FB. Postural motor learning in people with Parkinson's disease. *Journal of Neurology* 2016;**263**(8):1518–29.
14. Jacobs JV, Horak FB. Cortical control of postural responses. *Journal of Neural Transmission* 2007;**114**(10):1339–48.
15. Belen'kii VE, Gurfinkel VS, Pal'tsev EI. Control elements of voluntary movements. *Biofizika* 1967;**12**(1):135–41.
16. Mancini M, Zampieri C, Carlson-Kuhta P, Chiari L, Horak FB. Anticipatory postural adjustments prior to step initiation are hypometric in untreated Parkinson's disease: an accelerometer-based approach. *European Journal of Neurology* 2009;**16**(9):1028–34.
17. Cordo PJ, Nashner LM. Properties of postural adjustments associated with rapid arm movements. *Journal of Neurophysiology* 1982;**47**(2):287–302.
18. Frank JS, Patla AE. Balance and mobility challenges in older adults: implications for preserving community mobility. *American Journal of Preventive Medicine* 2003;**25**(3 Suppl. 2): 157–63.
19. Shimba T. An estimation of center of gravity from force platform data. *Journal of Biomechanics* 1984;**17**(1):53–60.
20. Winter DA. *The biomechanics and motor control of human gait: normal, elderly and pathological*. Waterloo (Ontario): Waterloo Biomechanics; 1991.
21. Winter DA, Ruder GK, MacKinnon CD. Control of balance of upper body during gait. In: Winters JM, Woo SLY, editors. *Multiple muscle systems: biomechanics and movement organization*. New York, NY: Springer New York; 1990. p. 534–41.
22. Winter DA, MacKinnon CD, Ruder GK, Wieman C. An integrated EMG/biomechanical model of upper body balance and posture during human gait. *Progress in Brain Research* 1993;**97**:359–67.
23. Cummings SR, Nevitt MC. Falls. *New England Journal of Medicine* 1994;**331**(13):872–3.
24. Sherrington CS. Flexion-reflex of the limb, crossed extension-reflex, and reflex stepping and standing. *The Journal of Physiology* 1910;**40**(1–2):28–121.
25. Diener HC, Ackermann H, Dichgans J, Guschlbauer B. Medium- and long-latency responses to displacements of the ankle joint in patients with spinal and central lesions. *Electroencephalography and Clinical Neurophysiology* 1985;**60**(5):407–16.
26. Takakusaki K. Functional neuroanatomy for posture and gait control. *Journal of Movement Disorder* 2017;**10**(1):1–17.
27. Jahn K, Zwergal A. Imaging supraspinal locomotor control in balance disorders. *Restorative Neurology and Neuroscience* 2010;**28**(1):105–14.
28. Jacobs JV, Lou JS, Kraakevik JA, Horak FB. The supplementary motor area contributes to the timing of the anticipatory postural adjustment during step initiation in participants with and without Parkinson's disease. *Neuroscience* 2009;**164**(2):877–85.
29. Cullen KE. Physiology of central pathways. *Handbook of Clinical Neurology* 2016;**137**:17–40.
30. Drew T, Dubuc R, Rossignol S. Discharge patterns of reticulospinal and other reticular neurons in chronic, unrestrained cats walking on a treadmill. *Journal of Neurophysiology* 1986;**55**(2):375–401.
31. Prentice SD, Drew T. Contributions of the reticulospinal system to the postural adjustments occurring during voluntary gait modifications. *Journal of Neurophysiology* 2001; **85**(2):679–98.
32. Gilat M, Shine JM, Walton CC, O'Callaghan C, Hall JM, Lewis SJG. Brain activation underlying turning in Parkinson's disease patients with and without freezing of gait: a virtual reality fMRI study. *NPJ Parkinson's Disease* 2015;**1**:15020.
33. Boisgontier MP, Cheval B, Chalavi S, et al. Individual differences in brainstem and basal ganglia structure predict postural control and balance loss in young and older adults. *Neurobiology of Aging* 2017;**50**:47–59.

34. Burciu RG, Fritsche N, Granert O, et al. Brain changes associated with postural training in patients with cerebellar degeneration: a voxel-based morphometry study. *Journal of Neuroscience: The Official Journal of the Society for Neuroscience* 2013;**33**(10):4594–604.
35. Mihara M, Miyai I, Hatakenaka M, Kubota K, Sakoda S. Role of the prefrontal cortex in human balance control. *NeuroImage* 2008;**43**(2):329–36.
36. Slobounov S, Hallett M, Stanhope S, Shibasaki H. Role of cerebral cortex in human postural control: an EEG study. *Clinical Neurophysiology: Official Journal of the International Federation of Clinical Neurophysiology* 2005;**116**(2):315–23.
37. Slobounov S, Wu T, Hallett M. Neural basis subserving the detection of postural instability: an fMRI study. *Motor Control* 2006;**10**(1):69–89.
38. Taubert M, Mehnert J, Pleger B, Villringer A. Rapid and specific gray matter changes in M1 induced by balance training. *NeuroImage* 2016;**133**:399–407.
39. Ouchi Y, Okada H, Yoshikawa E, Nobezawa S, Futatsubashi M. Brain activation during maintenance of standing postures in humans. *Brain: A Journal of Neurology* 1999;**122**(Pt 2):329–38.
40. Visser JE, Bloem BR. Role of the basal ganglia in balance control. *Neural Plasticity* 2005; **12**(2–3):161–74. discussion 263-72.
41. Drijkoningen D, Leunissen I, Caeyenberghs K, et al. Regional volumes in brain stem and cerebellum are associated with postural impairments in young brain-injured patients. *Human Brain Mapping* 2015;**36**(12):4897–909.
42. Honeycutt CF, Gottschall JS, Nichols TR. Electromyographic responses from the hindlimb muscles of the decerebrate cat to horizontal support surface perturbations. *Journal of Neurophysiology* 2009;**101**(6):2751–61.
43. Karachi C, Grabli D, Bernard FA, et al. Cholinergic mesencephalic neurons are involved in gait and postural disorders in Parkinson disease. *Journal of Clinical Investigation* 2010; **120**(8):2745–54.
44. Bakker M, De Lange FP, Helmich RC, Scheeringa R, Bloem BR, Toni I. Cerebral correlates of motor imagery of normal and precision gait. *NeuroImage* 2008;**41**(3):998–1010.
45. Bakker M, de Lange FP, Stevens JA, Toni I, Bloem BR. Motor imagery of gait: a quantitative approach. *Experimental Brain Research* 2007;**179**(3):497–504.
46. Jahn K, Deutschlander A, Stephan T, Strupp M, Wiesmann M, Brandt T. Brain activation patterns during imagined stance and locomotion in functional magnetic resonance imaging. *NeuroImage* 2004;**22**(4):1722–31.
47. Zwergal A, Linn J, Xiong G, Brandt T, Strupp M, Jahn K. Aging of human supraspinal locomotor and postural control in fMRI. *Neurobiology of Aging* 2012;**33**(6): 1073–84.
48. Solis-Escalante T, van der Cruijsen J, de Kam D, van Kordelaar J, Weerdesteyn V, Schouten AC. Cortical dynamics during preparation and execution of reactive balance responses with distinct postural demands. *NeuroImage* 2018;**188**:557–71.
49. Horak FB, Anderson ME. Influence of globus pallidus on arm movements in monkeys. I. Effects of kainic acid-induced lesions. *Journal of Neurophysiology* 1984;**52**(2):290–304.
50. Horak FB, Anderson ME. Influence of globus pallidus on arm movements in monkeys. II. Effects of stimulation. *Journal of Neurophysiology* 1984;**52**(2):305–22.
51. Wu T, Hallett M, Chan P. Motor automaticity in Parkinson's disease. *Neurobiology of Disease* 2015;**82**:226–34.
52. Chong RK, Horak FB, Woollacott MH. Parkinson's disease impairs the ability to change set quickly. *Journal of the Neurological Sciences* 2000;**175**(1):57–70.
53. MacKinnon CD. Sensorimotor anatomy of gait, balance, and falls. *Handbook of Clinical Neurology* 2018;**159**:3–26.
54. Buford JA, Davidson AG. Movement-related and preparatory activity in the reticulospinal system of the monkey. *Experimental Brain Research* 2004;**159**(3):284–300.

55. Matsuyama K, Drew T. Vestibulospinal and reticulospinal neuronal activity during locomotion in the intact cat. I. Walking on a level surface. *Journal of Neurophysiology* 2000;**84**(5):2237–56.
56. Canedo A, Lamas JA. Pyramidal and corticospinal synaptic effects over reticulospinal neurones in the cat. *The Journal of Physiology* 1993;**463**:475–89.
57. He XW, Wu CP. Connections between pericruciate cortex and the medullary reticulospinal neurons in cat: an electrophysiological study. *Experimental Brain Research* 1985;**61**(1):109–16.
58. Jinnai K. Electrophysiological study on the corticoreticular projection neurons of the cat. *Brain Research* 1984;**291**(1):145–9.
59. Kably B, Drew T. Corticoreticular pathways in the cat. I. Projection patterns and collaterization. *Journal of Neurophysiology* 1998;**80**(1):389–405.
60. Papegaaij S, Taube W, Baudry S, Otten E, Hortobagyi T. Aging causes a reorganization of cortical and spinal control of posture. *Frontiers in Aging Neuroscience* 2014;**6**:28.
61. Dijkstra BW, Horak FB, Kamsma YP, Peterson DS. Older adults can improve compensatory stepping with repeated postural perturbations. *Frontiers in Aging Neuroscience* 2015;**7**:201.

CHAPTER

2

Why is balance so important in Parkinson disease?

A. How common are balance disorders in PD?

Balance impairment is a frequent problem in people with Parkinson disease (PD). This fact has made some clinicians suggest that the PD triad of rest tremor, rigidity, and bradykinesia should be converted to a tetrad including balance impairment as a fourth feature. This fact indicates that impaired balance is an integral aspect of PD. The clinician can predict with confidence that almost every patient with PD will encounter problems with balance during the course of the disease and balance control will worsen with disease progression.

B. How important are balance and gait to the wellbeing of a person with PD?

The ability to safely ambulate is critical for good quality of life. The quality of life instrument, the PDQ-39,[1] was developed by investigators using focus groups where people with PD identified aspects of parkinsonism that detracted from their quality of life. Ten of the 39 items in the resultant scale relate to mobility, testifying to the importance of mobility to patients. Regression models examining the contribution of the Unified Parkinson Disease Rating Scale (UPDRS) measures, demographics, and functional measures of motor deficits to quality of life (PDQ-39) found that rigidity and tremor were unimportant. Functional measures of gait and bradykinesia (plus duration of disease) were the most important predictors of quality of life.[2]

Balance Dysfunction in Parkinson's Disease
https://doi.org/10.1016/B978-0-12-813874-8.00002-7

Although clinicians often focus on treating tremor, rigidity, and bradykinesia of arms and legs, balance and gait disturbances should receive more attention, as they impact quality of life more than limb disturbances.

The most obvious consequences of impaired balance are falls. Falls are estimated to be twice as common in PD compared to age-matched controls.[3] Patients with PD are over represented in elderly patients hospitalized for falls, have more severe injury, and longer hospital stays.[4] In fact, people with PD are four times more likely to have hip fractures than age matched controls.[3] The greater incidence of fractures with falls in people with PD suggests that people with PD fall poorly, without adequate rescue responses to avoid injuries. Complications of hip fractures in PD are more frequent, require longer hospital admissions, and are associated with poorer outcomes. Falls and the often accompanying fear of falling limits mobility and activity leading to further declines in mobility. But balance difficulties impact life in many other ways as well as predilection to falls. Balance is critical for walking and, in turn, affects performance of many daily activities around the home and in the community.

C. When do balance problems emerge?

Balance and gait difficulties are commonly thought to arise later in the disease. Indeed, the more advanced stages of disease in the Hoehn and Yahr scale[5] are defined by impairments in postural responses to external displacement, that is, a backward pull at the shoulders. The Hoehn and Yahr scale implies that balance problems are a late feature of the disease, appearing in the second half of the disease course. However, measuring turning during the Timed Up and Go test revealed slowing of turns very early in the disease, even when gait speed was normal[6] (Chapter 7). Sophisticated measures of postural sway during standing also reveals increased sway in early untreated people with PD[7] (Chapter 3). These findings are consistent with the clinical observation that falls occur in 80% of newly diagnosed people with PD followed for 4–5 years.[8] The clinical implication is that clinicians should be concerned about balance control as soon as people are diagnosed with PD.

D. Why does Parkinson disease affect balance?

Parkinson disease is classically thought of as a motor disorder and the motor symptoms are the basis of diagnosis. The motor problems of PD are related to the degeneration of the dopaminergic neurons in the substantia nigra, located in the midbrain. The dopaminergic neurons of the substantia nigra project to the basal ganglia which have an important role in learning motor sequences, choosing motor programs, vigor of movement, and automaticity of learned movements. Dopamine released by the substantial nigrafacilitates voluntary movement by the so called "direct pathway" which is stimulated by the dopamine D-1 receptor (Fig. 2.1).

Dopamine also suppresses unrelated movements via the so called "indirect pathway" which is inhibited by the dopamine D-2 receptor. Loss of the modulatory effects of dopamine on the basal ganglia reduces the activation of the direct pathway and as a consequence, reduces voluntary movement. In the indirect pathway, the reduction in dopamine disinhibits the indirect pathway, which increases the inhibition of the globus pallidus externa (GPe). This action then reduces the inhibitory output of the GPe to the excitatory subthalamic nucleus (STN) which projects to the GPi and increases the GPi inhibition of the thalamus. The inhibition of the thalamus VA/VL nuclei reduces excitatory output to the cortex. Although, the diagrams suggest that the excitatory dopaminergic dopamine D1 stimulation of the direct pathway and the D-2 inhibition of the indirect pathway would have opposing effects on movement, there is the evidence that the direct pathway leads to release of an intended movement and the indirect pathway suppresses other movements not related to the intended movement, that is, the indirect pathway acts like surround inhibition to prevent other superfluous movements. The loss of dopaminergic input to the striatum in PD will decrease the excitatory feedback to the cortex through the VA and VL nuclei of the thalamus to produce the classical triad of PD, rest tremor, rigidity, and bradykinesia.

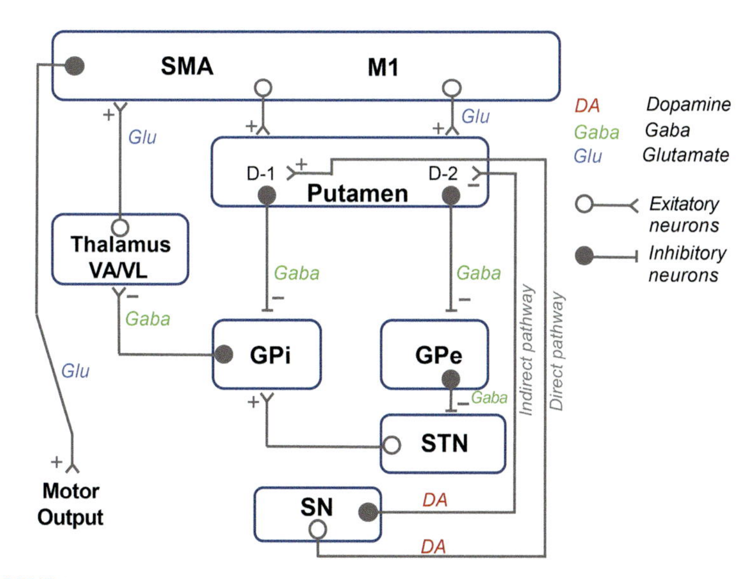

FIGURE 2.1 Schematic diagram of the basal ganglia–thalamocortical circuitry and their interactions. *M1*, Motor Areas; *SMA*, Supplementary Motor Area; *SN*, Substantia Nigra; *VA/VL*, Ventroanterior and Ventrolateral Nuclei; *GPi*, Globus Pallidus Internus; *GPe*, Globus Pallidus Externus; *STN*, Subthalamic Nucleus; Motor Output: Brainstem.

The characteristic loss of vigor of movement in PD is likewise seen in balance responses. Clinicians often talk of patients with PD "losing postural reflexes." However, loss of balance responses has never been demonstrated in studies of postural control. The postural responses are always present in people with PD and the onset is not delayed (Chapter 4). The postural responses are, however, slower to develop and the peak force and the responses may be ineffective because of excessive cocontraction of muscles.[9] That is, reactive postural responses and anticipatory postural adjustments and gait are hypometric, just like the limb movements that are characteristic of PD. This poor vigor in the execution of balance, locomotion, and voluntary limb movement can be seen in electromyographic recordings in PD. Repetitive but shorter, disorganized bursts of firing of agonist and antagonistic muscles occur in PD as opposed to the normal pattern with longer duration and synchronized agonist muscle burst followed by an antagonist burst in normal people.[10] Reversal of this abnormal firing pattern is the physiological basis for the predominant effect of dopamine on movement—reinstilling vigor to movements.[11] In addition, rigidity can produce stiffness of the axial segments that makes it difficult to turn and increases the velocity and jerkiness of sway (Chapters 3 and 7). Tremor may affect the legs, as well as arms, and a 4–6 Hz tremor is often apparent on a force platform on which a person with PD is standing.[12]

E. How does dopaminergic replacement therapy work?

Bradykinesia (slowed movement) and rigidity (increased muscle tone) are directly related to the extent of striatal dopaminergic denervation as measured by positron emission tomography (PET) with ^{18}F fluorodopa.[13] Parkinsonian tremor (4–6 Hz at rest) is also related to dopaminergic denervation, but the severity and extent of tremor are not proportional to dopaminergic denervation. Levodopa, the mainstay of medical treatment for PD, is intended to replenish the depleted basal ganglia dopamine. Levodopa is the immediate precursor of dopamine. It can cross the blood-brain barrier to enter the brain, unlike dopamine. In the brain, levodopa is converted to dopamine by a single enzymatic step. The dopamine derived from levodopa acts at the dopamine D-1 and D-2 receptors in the putamen to restore movement vigor; patients may appear almost normal early in the course of levodopa therapy. But, of course, orally administered levodopa cannot mimic the normal dopamine released from dopaminergic neurons that are firing in response to environment and internal signals. Another problem is that levodopa has a short half-life of one to 2 h requiring frequent doses of levodopa during the day. As parkinsonism progresses, the motor response to levodopa will begin to parallel the rise and fall of levodopa concentrations in the blood and dopamine concentrations in the brain. This produces a fluctuating motor response that is termed "ON-OFF" phenomenon when severe. The patient with PD may vacillate between minutes to hours of relatively good motor function interspersed with periods when the patient has marked restriction of movement. In addition, months to years of levodopa therapy will often produce unintended movements, levodopa-induced dyskinesia. Dyskinesia fluctuates during the day, accompanying "ON" time or when the effects of levodopa on motor function are evident. The dyskinesia can vary in severity; in some patients it is a subtle moving of the head or trunk or small movements in one hand. But in some people, the dyskinesia can be severe gyrations of the entire body. These two complications, motor fluctuations and dyskinesia, severely compromise the care of people with PD. The waxing and waning of the response to levodopa means that a single patient with PD will exhibit many different patterns of balance deficits during a single day.

Some, but not all, balance disorders respond to dopaminergic therapy and some aspect of balance control are made worse by levodopa (see levodopa section in Chapters 3–8). Locomotion, particularly straight ahead, and unimpeded walking are generally improved. The ability of levodopa to reduce balance difficulties is greatest in the earlier stages of PD. Eventually, most people with PD will develop abnormalities of balance and freezing of gait that are partially or not responsive to dopaminergic stimulation while the effects on limb rigidity and bradykinesia are preserved.

The dissociation between measures of parkinsonism in limbs and effects on balance may be explained by the anatomy of motor control in the distal limb muscles for dexterous movements versus the axial muscles critical to balance. Distal limb control is mediated by the lateral corticospinal tract (pyramidal tract). The proximal and axial muscles are controlled by the uncrossed portion of the pyramidal tracts (ventral corticospinal tracts), but also the tectospinal, vestibulospinal, and reticulospinal tracts arising in the brainstem that bilaterally innervate the proximal and axial muscles. These different pathways may explain why levodopa replacement therapy improves rigidity of the limbs but not axial rigidity.[14] It is important to recognize that some aspects of balance and locomotion may be worsened by levodopa, partially because of levodopa-induced dyskinesia.[15] Levodopa-induced dyskinesia, affects standing balance, execution of postural responses, and postural transitions, although the effects of dyskinesia have not been fully explored yet. But with progression of PD pathology, postural synergies themselves are affected by pathology outside the substantia nigra. This results in distorted postural responses and an inability to adapt to the environmental demands.

Although a very different modality, deep brain stimulation (DBS) of the GPi and STN may improve the same clinical features of parkinsonism that are improved by levodopa. If the parkinsonian features do not respond to levodopa, albeit briefly, DBS is unlikely to improve the clinical feature. Thus DBS can improve gait speed and standing postural control as does levodopa. But DBS often worsens dynamic postural control and may increase falls and freezing of gait[16] (see DBS sections in Chapters 3–8).

F. Do dopaminergic mechanisms explain all the balance deficits of parkinsonism?

Although the loss of the dopamine neurons of the substantia nigra is the sine qua non of PD, the pathology of PD is much more widespread than just the substantia nigra. According to Braak's staging of the pathology of PD,[17] PD first begins in peripheral nervous system, autonomic nervous system, and/or central nervous system (dorsal motor nucleus of the vagal nerve). At stage 2, the pathology spreads to the raphe (serotoninergic) and locus coeruleus (noradrenergic) neurons. In stage 3, the substantia nigra (dopaminergic) and the pedunculopontine nuclei[18] (cholinergic, gabaergic, and glutaminergic) as well as the spinal cord and amygdala are affected. In stage 4, the thalamus and temporal lobe may have pathology. In stage 5, sensory association areas and prefrontal regions show pathology. In the final, stage 6, primary motor and sensory areas are also affected. Therefore, balance may be affected by multiple mechanisms with ever expanding pathology during the progression of the disease. Balance dysfunction will be much more than just hypometric responses. The most common pathological stage of PD in over 2300 autopsies is stage 3,[17] when the substantia nigra and pedunculopontine nuclei (PPN) are affected, in addition to the serotonergic and noradrenergic nuclei of the brainstem affected in stage 2. These observations emphasize the importance of not just dopaminergic but cholinergic, serotonergic, noradrenergic, and conceivably glutaminergic and gabaergic mechanisms in midstage PD.

The cholinergic system is the other neurotransmitter to receive much clinical attention in PD at this time. Cholinergic innervation of the hemispheres emanates from the midbrain and projects to all cortical areas. This cortical cholinergic innervation has been linked to cognition and gait speed.[19] Cholinergic innervation of the thalamus mainly arises from the PPN, a nucleus that arguably corresponds to the midbrain locomotor area which controls postural tone required for locomotion as well as locomotion itself.[20] Loss of cholinergic input to the thalamus has been linked to falls in PD,[21] and there is some evidence for efficacy of cholinergic drugs for balance or preventing falls.[22,23] Conversely, anticholinergic drugs have been linked to falls in the elderly, further evidence of the importance of cholinergic system in balance.[24,25]

Beyond monoaminergic and cholinergic systems, what about other brain regions and circuits? Gait, as defined by alternating stepping, is largely programmed at the spinal level and is turned ON and OFF by descending brainstem influences from brainstem locomotor regions. Cortical areas modify gait to environmental demands and to a person's goals. Control of balance is dependent upon recognition of where the body is in space and in the gravitational field, as well as attention to the pertinent aspects of environment. These functions depend upon cortical circuits. Thus pathology in cortical areas in PD, generally related to deposition of alpha synuclein in cortical neurons in addition to decreased dopaminergic projections to frontal regions, underlies some of the balance dysfunction found in PD. In addition, presumed small vessel disease leads to subcortical white matter hyperintensities and other vascular abnormalities with aging. These lesions are superimposed on the pathology of PD and may contribute to abnormalities of balance.[26]

The spread of the PD pathology to cortical regions and to a multitude of neurotransmitter systems and cortical circuits gives rise to the complex pattern of balance deficits that may occur in PD. All patients with idiopathic PD will have dopamine-sensitive, progressive bradykinesia and rigidity that impair balance control. But every patient with PD, even those who are "tremor dominant," thought to be a more benign form of PD, will eventually suffer balance disorders. However, each individual with PD will experience different combinations of balance impairments. The balance impairments in an individual will depend on their preexisting or concomitant comorbidities affecting balance (i.e., musculoskeletal constraints, sensory loss, cortical small vessel disease, etc.), pattern of dopamine neuron loss, and the cortical circuits affected. In addition, some subjects will have other complications of PD including impaired kinesthesia, cognitive impairment, and mood changes affecting balance. Others will experience freezing of gait and concomitant anxiety that result in fear of falling and distinctive balance impairments. Differences in the constellation of balance impairments for each patient mean it is critical to customize treatment based on a comprehensive balance assessment (Fig. 2.2).

G. How is balance affected in Parkinson-Plus syndromes?

Parkinson disease is the most common cause of bradykinesia and rigidity, but these signs, coupled with balance impairments, may occur in pathological entities that may be difficult to separate from idiopathic PD,

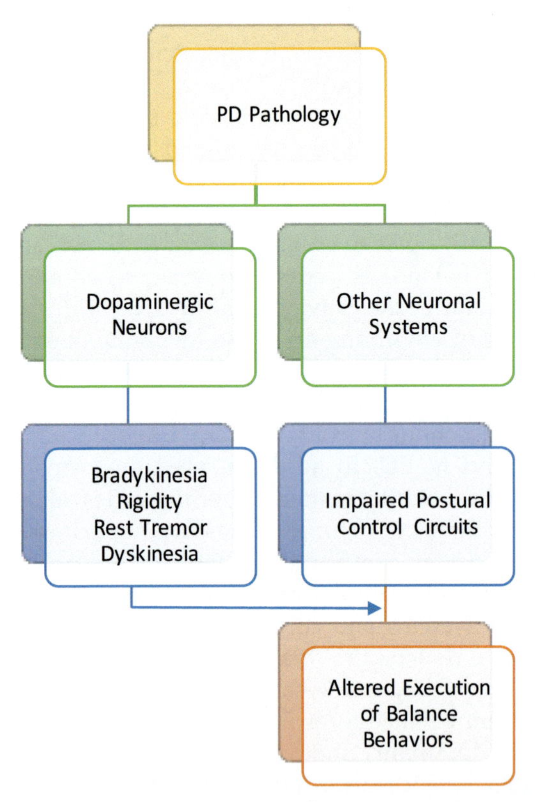

FIGURE 2.2 Parkinson disease (Alpha Synuclein) pathology in the dopaminergic system causing bradykinesia, rigidity, rest tremor, and dyskinesia can affect the execution of balance synergies. However, PD pathology in other neurotransmitter systems and circuits affects balance mechanisms so that the clinical picture is a summation of dopaminergic and other circuit deficits.

particularly early in the disease course.[27] The two most commonly encountered atypical parkinsonian disorders are multiple system atrophy (MSA) and progressive supranuclear palsy (PSP). In these disorders, in addition to degeneration of the dopaminergic system, there are pathological changes in other hemispheric and brainstem regions contributing to the balance difficulties in these patients. MSA and PSP frequently have more severe balance difficulties and earlier falls in the course of the disease than idiopathic PD. And early onset of imbalance and falls should make these disorders a consideration. Unlike idiopathic PD, balance problems in MSA may also result from cerebellar ataxia and spasticity, as well as parkinsonism. PSP is associated with frontal lobe pathology that affects voluntary gaze and behavior. Falls occur early in the course of the disease and are often backwards. The frontal lobe pathology frequently causes disinhibition which may make gait reckless as the patient does not appreciate their balance impairment and fall risk. So called vascular parkinsonism is generally related to cerebral small vessel disease affecting white matter, particularly in the frontal lobe. Vascular parkinsonism is sometimes called lower body parkinsonism because it affects gait, turning, and freezing similarly to idiopathic PD, although the upper body can be relatively spared. That is, people with vascular parkinsonism have normal arm swing, no resting tremor, but shuffling gait and freezing of gait. Unlike idiopathic PD, they tend to have wide-based walking.[28] In addition, vascular disease and particularly small vessel disease in the frontal and periventricular brain areas in idiopathic PD may contribute to their balance problems.

H. Do nonmotor signs affect balance control in PD?

Although motor dysfunction is the aspect of PD on which the diagnosis of PD is based, PD involves much more than just motor dysfunction. Nonmotor problems are extensive and implicate much wider neurologic dysfunction.[29] Five nonmotor features that may contribute to balance impairments are cognitive impairment, sensory dysfunction, anxiety, depression, and autonomic insufficiency. Cognitive changes tend to be mild cognitive impairment with executive dysfunction but cognitive changes may be extensive and be frank dementia. Cognitive changes may affect balance in several ways and are often associated with balance changes as the disease progresses because of shared putative central mechanisms, as will be explored in following chapters. Anxiety is common in PD, especially in those with freezing of gait,[30] and may affect postural control and gait.[31] Depression is associated with slowing of movement and gait[32] that likely also affects balance control. Sensory alterations may have both peripheral and central etiologies in PD and distort spatial sensory integration which is essential for balance.[33,34] Orthostatic hypotension may be a cause of falls, often collapsing falls with loss of muscle tone that must be differentiated from other impairments of balance. This is an important etiology of falls to be aware of as the condition generally responds to treatment of orthostatic blood pressure drops. In summary, balance is more than just motor dysfunction; sensory, cognitive, affective, and autonomic dysfunction, all of which are common in PD, and contribute to the balance dysfunction of PD (Fig. 2.3).

Balance, and the ability to avoid falls while maintaining mobility, is dependent upon multiple aspects of brain control of balance, almost all of which may be affected by PD. In the chapters that follow, the abnormalities in standing balance, responses to perturbations, anticipatory balance adjustments, balance while walking, freezing, and turning will be addressed. The effects of PD treatments on balance strategies will be described and approaches to remedying impaired mobility explored.

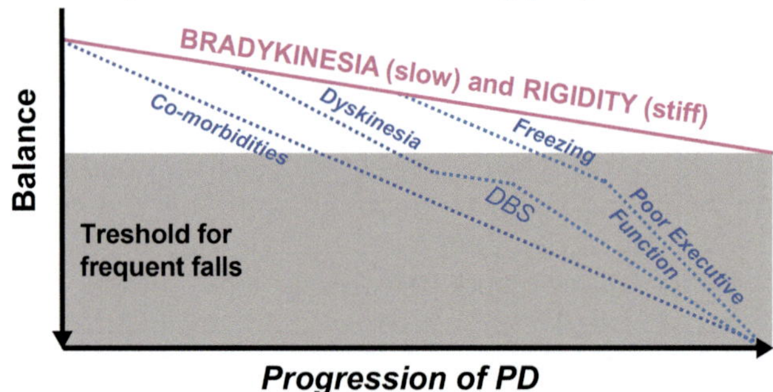

FIGURE 2.3 Progression of imbalance in PD until they reach the threshold for frequent falls as the disease progresses. Dopaminergic loss (pink line) in Parkinson disease and other disease features (blue dotted lines) that contribute to imbalance and alter progression of PD.

Highlights

- Every person with PD will have balance difficulties during the course of the disease.
- Balance and gait abnormalities appear early in the disease.
- Balance is essential for mobility, and is very important to quality of life, more so than limb function.
- Falls are more common, more severe, and more life threatening in PD than other neurological diseases.
- Balance disturbances in Parkinson disease include multiple aspects of balance including standing balance, reactive postural responses, anticipatory postural adjustments prior to gait initiation, and dynamic balance during walking and turning.
- Balance disturbances are more than just a manifestation of motor dysfunction in PD. Cognitive, affective, sensory, and autonomic dysfunctions contribute to balance disorders in PD.
- Levodopa and DBS have both positive and negative effects on balance.

References

1. Jenkinson C, Fitzpatrick R, Peto V, Greenhall R, Hyman N. The Parkinson's Disease Questionnaire (PDQ-39): development and validation of a Parkinson's disease summary index score. *Age and Ageing* 1997;**26**(5):353–7.
2. Ellis T, Cavanaugh JT, Earhart GM, Ford MP, Foreman KB, Dibble LE. Which measures of physical function and motor impairment best predict quality of life in Parkinson's disease? *Parkinsonism & Related Disorders* 2011;**17**(9):693–7.

3. Kalilani L, Asgharnejad M, Palokangas T, Durgin T. Comparing the incidence of falls/fractures in Parkinson's disease patients in the US population. *PLoS One* 2016;**11**(9):e0161689.
4. Paul SS, Harvey L, Canning CG, et al. Fall-related hospitalization in people with Parkinson's disease. *European Journal of Neurology* 2017;**24**(3):523–9.
5. Hoehn MM, Yahr MD. Parkinsonism: onset, progression and mortality. *Neurology* 1967;**17**(5):427–42.
6. Zampieri C, Salarian A, Carlson-Kuhta P, Aminian K, Nutt JG, Horak FB. The instrumented timed up and go test: potential outcome measure for disease modifying therapies in Parkinson's disease. *Journal of Neurology Neurosurgery and Psychiatry* 2010;**81**(2):171–6.
7. Mancini M, Horak FB, Zampieri C, Carlson-Kuhta P, Nutt JG, Chiari L. Trunk accelerometry reveals postural instability in untreated Parkinson's disease. *Parkinsonism & Related Disorders* 2011;**17**(7):557–62.
8. Lord S, Galna B, Yarnall AJ, Coleman S, Burn D, Rochester L. Predicting first fall in newly diagnosed Parkinson's disease: insights from a fall-naive cohort. *Movement Disorders: Official Journal of the Movement Disorder Society* 2016;**31**(12):1829–36.
9. Horak FB, Nutt JG, Nashner LM. Postural inflexibility in parkinsonian subjects. *Journal of the Neurological Sciences* 1992;**111**(1):46–58.
10. Hallett M, Khoshbin S. A physiological mechanism of bradykinesia. *Brain: A Journal of Neurology* 1980;**103**(2):301–14.
11. Albin RL, Leventhal DK. The missing, the short, and the long: levodopa responses and dopamine actions. *Annals of Neurology* 2017;**82**(1):4–19.
12. Kerr G, Morrison S, Silburn P. Coupling between limb tremor and postural sway in Parkinson's disease. *Movement Disorders: Official Journal of the Movement Disorder Society* 2008;**23**(3):386–94.
13. Pikstra ARA, van der Hoorn A, Leenders KL, de Jong BM. Relation of 18-F-Dopa PET with hypokinesia-rigidity, tremor and freezing in Parkinson's disease. *NeuroImage Clinical* 2016;**11**:68–72.
14. Wright WG, Gurfinkel VS, Nutt J, Horak FB, Cordo PJ. Axial hypertonicity in Parkinson's disease: direct measurements of trunk and hip torque. *Experimental Neurology* 2007;**208**(1):38–46.
15. Curtze C, Nutt JG, Carlson-Kuhta P, Mancini M, Horak FB. Levodopa is a double-edged sword for balance and gait in people with Parkinson's disease. *Movement Disorders: Official Journal of the Movement Disorder Society* 2015;**30**(10):1361–70.
16. Collomb-Clerc A, Welter ML. Effects of deep brain stimulation on balance and gait in patients with Parkinson's disease: a systematic neurophysiological review. *Neurophysiologie Clinique* 2015;**45**(4–5):371–88.
17. Del Tredici K, Braak H. Review: sporadic Parkinson's disease: development and distribution of alpha-synuclein pathology. *Neuropathology and Applied Neurobiology* 2016;**42**(1):33–50.
18. Pienaar IS, Vernon A, Winn P. The cellular diversity of the pedunculopontine nucleus: relevance to behavior in health and aspects of Parkinson's disease. *The Neuroscientist* 2017;**23**(4):415–31.
19. Bohnen NI, Frey KA, Studenski S, et al. Gait speed in Parkinson disease correlates with cholinergic degeneration. *Neurology* 2013;**81**(18):1611–6.
20. Takakusaki K. Functional neuroanatomy for posture and gait control. *Journal of movement disorders* 2017;**10**(1):1–17.
21. Bohnen NI, Muller ML, Koeppe RA, et al. History of falls in Parkinson disease is associated with reduced cholinergic activity. *Neurology* 2009;**73**(20):1670–6.
22. Chung KA, Lobb BM, Nutt JG, Horak FB. Effects of a central cholinesterase inhibitor on reducing falls in Parkinson disease. *Neurology* 2010;**75**(14):1263–9.

23. Henderson EJ, Lord SR, Brodie MA, et al. Rivastigmine for gait stability in patients with Parkinson's disease (ReSPonD): a randomised, double-blind, placebo-controlled, phase 2 trial. *The Lancet Neurology* 2016;**15**(3):249–58.
24. Cardwell K, Hughes CM, Ryan C. The association between anticholinergic medication burden and health related outcomes in the 'oldest old': a systematic review of the literature. *Drugs & Aging* 2015;**32**(10):835–48.
25. Chatterjee S, Bali V, Carnahan RM, Chen H, Johnson ML, Aparasu RR. Anticholinergic medication use and risk of fracture in elderly adults with depression. *Journal of the American Geriatrics Society* 2016;**64**(7):1492–7.
26. Vesely B, Antonini A, Rektor I. The contribution of white matter lesions to Parkinson's disease motor and gait symptoms: a critical review of the literature. *Journal of Neural Transmission* 2016;**123**(3):241–50.
27. Bhidayasiri R, Rattanachaisit W, Phokaewvarangkul O, Lim TT, Fernandez HH. Exploring bedside clinical features of parkinsonism: a focus on differential diagnosis. *Parkinsonism & Related Disorders* 2018;**59**:74–81.
28. Fling BW, Dale ML, Curtze C, Smulders K, Nutt JG, Horak FB. Associations between mobility, cognition and callosal integrity in people with parkinsonism. *NeuroImage Clinical* 2016;**11**:415–22.
29. Chaudhuri KR, Schapira AH. Non-motor symptoms of Parkinson's disease: dopaminergic pathophysiology and treatment. *The Lancet Neurology* 2009;**8**(5):464–74.
30. Martens KAE, Hall JM, Gilat M, Georgiades MJ, Walton CC, Lewis SJG. Anxiety is associated with freezing of gait and attentional set-shifting in Parkinson's disease: a new perspective for early intervention. *Gait & Posture* 2016;**49**:431–6.
31. Doumas M, Morsanyi K, Young WR. Cognitively and socially induced stress affects postural control. *Experimental Brain Research* 2018;**236**(1):305–14.
32. Pirker W, Katzenschlager R. Gait disorders in adults and the elderly: a clinical guide. *Wiener Klinische Wochenschrift* 2017;**129**(3–4):81–95.
33. Zis P, Grunewald RA, Chaudhuri RK, Hadjivassiliou M. Peripheral neuropathy in idiopathic Parkinson's disease: a systematic review. *Journal of the Neurological Sciences* 2017;**378**:204–9.
34. Sailer A, Molnar GF, Paradiso G, Gunraj CA, Lang AE, Chen R. Short and long latency afferent inhibition in Parkinson's disease. *Brain: A Journal of Neurology* 2003;**126**(Pt 8):1883–94.

CHAPTER

3

How is balance during quiet stance affected by PD?

Clinical case

When Tom was diagnosed with Parkinson disease (PD), he felt his balance was good while doing activities around the house. However, when his physical therapist did a posturography assessment he noticed that his postural sway while standing was very irregular and his limits of stability were reduced. Levodopa replacement helped his standing balance for a few years. However, later on, he developed Pisa syndrome with leaning to the right side, and levodopa gave him dyskinesia. In addition, his neck felt very rigid and turning mobility also became difficult. His doctor suggested deep brain stimulation (DBS) in the internal Globus Pallidus (GPi), as it is more effective for balance impairments than DBS in the STN, and may improve flexed postural alignment. However, Tom read about a new exercise program focused on balance exercises and decided to try that first for 6 weeks. Improvements in his balance, limits of stability, neck rigidity, and postural alignment were excellent.

A. How does PD affect postural alignment?

As PD progresses, patients stand with an increasingly stooped posture with rounding of the shoulders and flexion of the hips and knees, reflecting increased flexor tone.[1] This classic, stooped posture is the most common postural deformity observed in individuals with PD. Interestingly, when healthy people voluntarily assume this flexed posture, their postural stability, especially in the backward direction, becomes impaired[2,3] (see Chapter 4). The etiology of the postural misalignment in

https://doi.org/10.1016/B978-0-12-813874-8.00003-9

PD is not clear. Background muscle activation is larger than normal in people with PD, especially in flexor muscles, which could contribute to their flexed posture.[1] A tilted or inaccurate internal representation of postural or visual verticality could also result in postural alignment that is not aligned with gravity.[4,5] An altered sense of postural verticality may relate to impaired proprioception and affect the position of the body center of mass over the base of support, making patients more vulnerable to falls.[6,7]

In addition to the characteristic flexed posture (Fig. 3.1A), up to one-third of patients with PD have deformities of their neck or trunk that may include camptocormia (forward bent trunk; Fig. 3.1B), anterocollis (flexed neck), Pisa syndrome (lateral lean of trunk, Fig. 3.1C), and scoliosis.[8–12] As reviewed by Doherty et al., many etiologies may exist for these other trunk abnormalities, including dystonia, muscle weakness, and orthopedic abnormalities. These postural disorders are often associated with musculoskeletal pain.[10] Since the basal ganglia project directly to the tonogenic centers in the brainstem, such as the pedunculopontine nucleus, abnormally high and/or asymmetrical postural muscle tone is likely involved in abnormal postural alignment in people with PD.

Camptocormia is an axial postural deformity characterized by abnormal thoracolumbar spinal flexion. The symptom usually presents while standing, walking, or exercising, and is alleviated while sitting, lying in a recumbent position, standing against a wall, or using walking

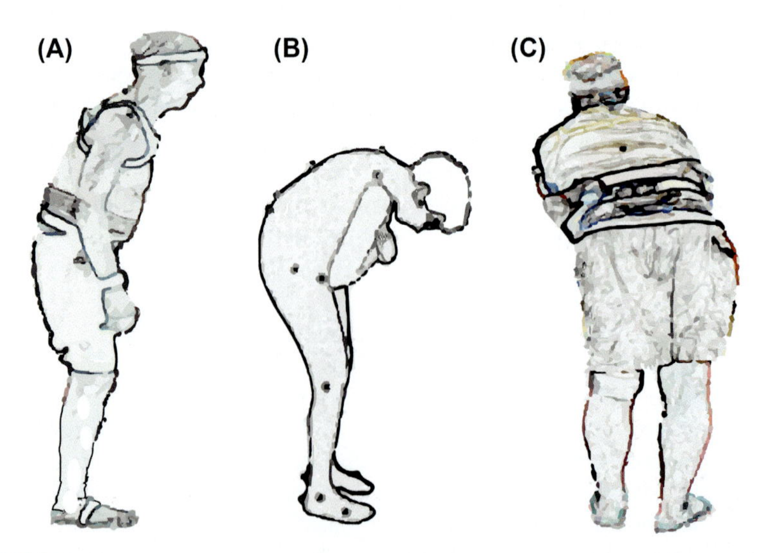

FIGURE 3.1 Patients with stooped posture (A), Camptocormia (B), and Pisa syndrome with camptocormia (C). Subjects are wearing reflective markers to quantify their postural aliglnment.

support. There is no consensus on the degree of thoracolumbar flexion to define camptocormia. However, most authors usually use an arbitrary number of at least 45 degrees flexion of the thoracolumbar spine when the individual is standing or walking. However, the camptocormia develops over seconds to hours and the degree of flexion depends upon the time the patient has been standing and what sensory aids the patient is using.[13,14] Etiologies of camptocormia are heterogeneous, and PD is one of its many causes. Mainly, the pathogenesis of camptocormia can generally be divided into two areas: a peripheral disorder that causes weakness in extensor postural muscles[15–19] or a central disorder that disrupts the tonic central drive to the postural muscles.[20–23] The prevalence of camptocormia in PD ranges from 3% to 18%. Although there is no established consensus for treatment of camptocormia in PD, there are nonpharmacological, pharmacological, and surgical approaches that can be used (see paragraphs below).

Pisa syndrome was first described in 1972 in patients treated with neuroleptics. Since 2003, when it was first reported in patients with PD, Pisa syndrome has progressively drawn the attention of clinicians and researchers.[12] Pisa syndrome is clinically defined as lateral truncal flexion of $\geq$10 degrees, that is not due to an underlying mechanical restriction and is almost completely alleviated by passive mobilization or supine positioning.[10] Clinically, Pisa can be classified as mild (<20 degrees) or severe (>20 degrees).[12] In addition, the leaning is only to one side and this laterality of deviation is preserved throughout their disease. Contrary to initial reports, Pisa syndrome is common among patients with PD, with an estimated prevalence of 8.8% according to a large survey.[12] Pisa syndrome is also associated with the following specific patient features: more severe motor phenotype, ongoing combined pharmacological treatment with levodopa and dopamine agonists, gait disorders, and such comorbidities as osteoporosis and arthrosis.[12,24–26] The present literature on treatment outcomes is scarce, and the uneven effectiveness of specific treatments has produced conflicting results. This might be because of the limited knowledge of Pisa syndrome pathophysiology and its variable clinical presentation, which further complicates designing randomized clinical trials on this condition. However, because some forms of Pisa syndrome are potentially reversible, there is growing consensus on the importance of its early recognition and the importance of pharmacological adjustment and rehabilitation.

Interestingly, both camptocormia's forward trunk flexion and Pisa syndrome's lateral trunk flexion are not always present, but develop gradually over many seconds to minutes when the person assumes an upright standing and/or sitting position. A concurrent dual task can increase the speed of developing a flexed posture, suggesting that patients lack automatic control of their postural alignment so they benefit from devoting conscious attention to posture.[14] Consistent with the benefits of conscious attention on posture, visual feedback from a mirror or inclinometer, aiming with an arm, and light touch can significantly improve postural alignment (see Fig. 3.2). These are commonly called "sensory tricks" and are characteristic of both camptocormia and Pisa syndrome.

Pisa syndrome can develop after exposure to various medications (neuroleptics and cholinesterase inhibitors) and also after the modification of antiparkinsonian medication regimens in patients with PD.[27] More commonly, Pisa presents as postural deviation in patients with various neurodegenerative disorders, including PD, independent of any medication exposure. There is currently no consensus on the pathophysiologic mechanisms driving the development of Pisa in PD; however, both central and peripheral etiologies have been implicated.[28] Although there are many hypotheses about the pathophysiologic mechanisms causing lateral truncal flexion in patients with Pisa and PD, the presence of multifocal central dysfunction has been most widely implicated. This "central generator" likely includes asymmetry of basal ganglia outflow to tonogenic brainstem centers[28–31] and abnormalities in spatial cognition which

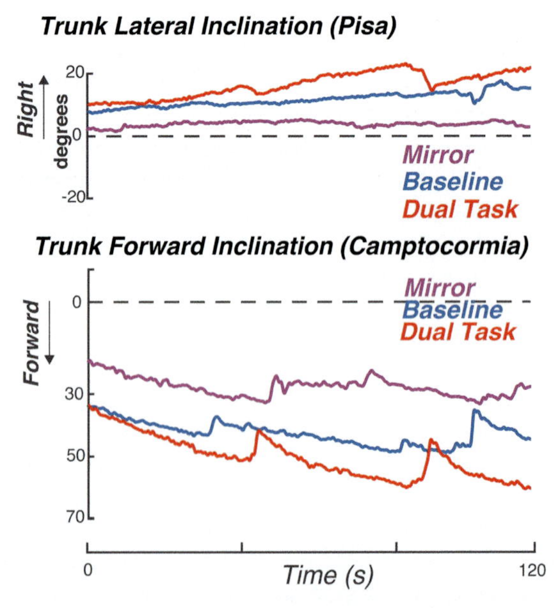

FIGURE 3.2 Progressive lateral (*upper panel*) and forward (*lower panel*) leaning in a representative subject with Pisa syndrome and camptocormia. The baseline condition (*blue*) shows gradually increasing lateral (*top*) and forward (*bottom*) trunk flexion over 2 min of standing. Adding a concurrent cognitive dual task (*red*) speeds up the trunk inclinations and providing visual feedback of posture with a mirror (*purple*) slows or eliminates the inclinations.

include dysfunctional sensorimotor (visual, proprioceptive, and vestibular) integration systems[32] and asymmetric vestibular dysfunction.[33]

The development of Pisa in PD can cause a significant increase in disability and there is currently no effective evidence-based therapies that significantly rectify lateral postural deviation in this syndrome.

B. How are limits of stability impaired in PD?

The limits of stability can be defined as the maximum, voluntary, inclined posture while standing.[34] Statically holding the body CoM and the CoP near the forward or backward limits of foot support simulates functional positions that occur in motor tasks such as in reaching and the transitions from stance-to-gait and from sit-to-stand. Limits of stability, quantified by the maximum, voluntary inclined posture may be considered "functional" limits of stability, since they are influenced by subjective perception, internal postural control abilities, and environmental factors, and not only by body biomechanics, muscle strength, or segment properties, such as size of the feet.[35] The extent to which a person can lean toward these biomechanical limits of stability is also affected by the person's confidence in their ability to return their center of body mass (CoM) over their base of support. Under dynamic conditions, the limits of stability have been defined as a ratio of the maximum displacement of the CoM with respect to the maximum displacement of the center of pressure (CoP) during a feet-in-place postural response to external perturbations that can be controlled without a fall or a step[36] (see also Chapter 4).

Patients with PD have smaller functional limits of stability in the sagittal plane compared to age-matched control subjects.[35] The small stability limits in subjects with PD is mainly due to reduced maximum forward (not backward) body leaning (Fig. 3.3). Potentially, this small maximum forward leaning could be related to their impaired postural preparation for gait initiation[37–39] that similarly requires a preparatory forward lean, but this relationship has not been investigated.[40]

In contrast to the forward direction, stability limits in the backward direction were not significantly different between control and PD subjects (Fig. 3.3). This result could be due to an age-effect or "floor"-effect on maximum backward inclination common to both PD and control subjects due to biomechanical constraints for backward leaning.[34]

People with PD maintain their stooped posture during voluntary leaning tasks.[2,35] Postural kinematic strategies[41] in the steady-state, upright and their maximum leaning positions confirm no change in the typical, stooped posture of subjects with PD.[3] The stooped posture probably contributes to their reduced forward limits of stability, because the flexed ankle, knee, and hip joints result in longer ankle plantar flexor

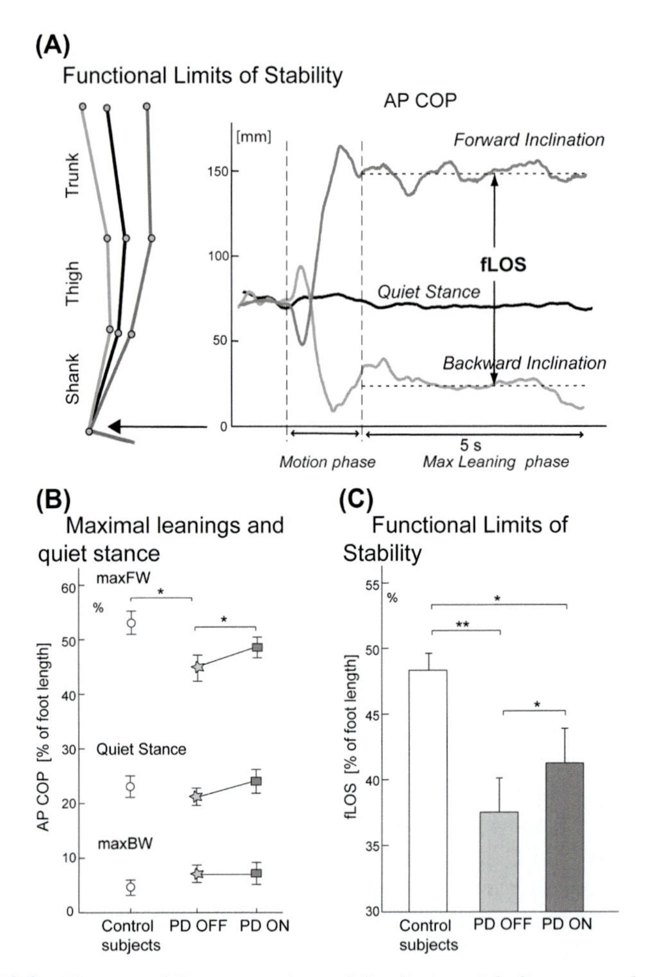

FIGURE 3.3 *Upper panel*: Representative stability limits, stick diagram, and CoP excursion during quiet stance and the motion and leaning phase of maximal forward and backward leaning directions in a healthy elderly subject. Zero in the y-axis represents the heel position. *Lower panel*: Comparison between means from a control group and PD group in the ON and OFF levodopa state for their (A) position of anteroposterior (AP) center of pressure (CoP mean and Standard Deviation, SD) during the maximal leaning tasks and in quiet stance; (B) functional limits of stability (fLOS) (mean and SD) quantified as the difference between maximal forward (maxFW) and maximal backward (maxBW) lean. (C) Limits of stability were larger (improved) with levodopa but were still significantly smaller than in age-matched control subjects. *From Mancini M, Rocchi L, Horak FB, Chiari L. Effects of Parkinson's disease and levodopa on functional limits of stability.* Clinical Biomechanics *2008;* **23***(4):450–8.*

muscles and larger antigravity forces required to maintain equilibrium. This unchanged body postural alignment across tasks is consistent with previous studies showing that subjects with PD have difficulty changing postural strategies with changes in initial conditions.[3,37,39,42]

C. What is the contribution of axial tone to standing balance?

Disturbance to the control of axial tone may underlie many of the gait and postural impairments found in PD. For example, axial hypertonicity is thought to contribute to the limitations on body rotation during sleep[43,44] and may also contribute to abnormal intersegmental coordination during walking and turning.[45–47]

A few direct measures of axial tone in PD have been reported. A bedside technique for measuring axial rigidity has been used to assess trunk tone by manipulating the legs and hips while the patient was supine.[48,49] In these two studies, subjects with PD had greater tone than control subjects, but a drawback of this technique is that in the supine position, the body is relaxed and postural tone is reduced, making this technique useful only when tone is very high. A more recent study quantified differences in trunk tone between PD and healthy individuals using a device that isokinetically flexed and extended the trunk.[50] Subjects with PD had greater trunk extension resistance compared to healthy control subjects, and to a lesser extent during trunk flexion, but only when the trunk was repositioned with high angular velocities (>60 degrees/s). However, neither the bedside nor the isokinetic technique isolated and measured hip torque; instead, these techniques focused on the trunk alone.

To measure more directly axial and proximal tone, Gurfinkel and colleagues[13] developed a novel device to torsionally twist the body about the vertical axis during natural standing posture. This device, called "Twister," quantifies axial tone by measuring resistance to passively applied torsional rotation at the neck, trunk, and/or hips without constraining anterior-posterior, lateral, or vertical body position or postural sway.[13] Using the twister device, it has been shown that axial tone in healthy adults is not static, but rather involves flexible, active shortening and lengthening reactions.[13] Unlike stretch reflexes, in which muscles are activated during muscle lengthening and silenced during muscle shortening, shortening reactions are associated with increased activity during muscle shortening and decreased activity during muscle lengthening.[51]

Using this twister methodology, subjects with PD had significantly more rigidity in the trunk, hips, and neck compared to healthy adults.[52] Fig. 3.4 *left panel* shows an example of larger neck torque changes during axial

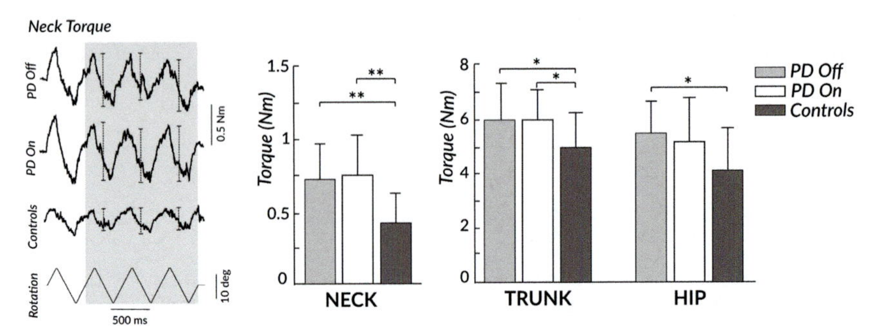

FIGURE 3.4 *Left panel*: Representative neck torque in a PD OFF and ON levodopa medication, and a healthy adult during slow rotations of the body when standing and the head fixed in space. Notice no change in neck torque ON versus OFF but larger than control subject. Shaded area shows period in which torque was measured for the histogram in right panel. *Right panel*: Mean and SD of neck, hip, and trunk torque in PD group compared to healthy control group during stance. *Adapted from Franzen E., et al. Reduced performance in balance, walking, and turning tasks is associated with increased neck tone in Parkinson's disease. Experimental Neurology. 2009, 430-438.*[53]

rotation of the body with the surface but with the head fixed to earth in a subject with PD OFF and ON compared to an age-matched subject without PD.[53] On average, neck tone is 75% higher in PD compared to controls, compared to 22% at the trunk and 32% at the hips (Fig. 3.4, *right panel*). In addition, the increase in neck tone was more strongly related to functional mobility performance (timed Figure 8 test and rolling) than were the increases in trunk and hip axial tone, suggesting that neck muscle tone plays a critical role in the control of balance, mobility, and coordination.[53]

Direct, objective measures of axial rigidity correlated with clinical measures of disease severity. These correlations suggest that axial rigidity reflects the overall postural tone, despite having tonogenic pathways dissociable from those projecting to limb musculature. Higher tone in the neck, than the trunk and hips, in subjects with PD may be related to the descending and ascending innervations of neck motoneurons.[52,53]

Many everyday motor tasks require intersegmental coordination of axial segments to maintain balance and mobility. Stability and mobility in functional motor activities depend on the precise regulation of axial phasic and tonic activity that is carried out automatically, without conscious awareness.[13] In PD, posture and gait movements involving axial counterrotation are often problematic. During walking and turning, individuals with PD tend to rotate the head and trunk simultaneously, whereas healthy subjects lead turns with prior head rotation.[45–47,54] Effective postural responses to recover equilibrium also require precise spatial-temporal coordination of the axis (Chapter 4).[55] In addition, control of sagittal stability requires coordinated counterrotation of the upper and lower body segments and lateral postural stability requires coordination of lateral trunk flexion with trunk rotation; coordinations that are disrupted by rigidity in PD.[36,56–58]

D. Why is postural sway important?

During quiet stance, the body CoM is located within a base of support defined by the feet. However, the body is not entirely still, there is continuous movement of the CoM termed "postural sway." Postural sway represents a state of complex sensorimotor control loops that contribute to balance control (as described in Chapter 1).[59] Postural sway can be measured via a force plate that detects fluctuations of the CoP, or by accelerometers that detect fluctuations of the body CoM. The variation in sway can be characterized by a number of variables such as sway area, velocity, frequency, and maximum direction of sway.

Depending on the size of the base of support, more or less postural sway can be tolerated. For example, less sway of the body CoM is tolerated when attempting to balance on one foot compared to standing with the two feet apart. Sway area, velocity, and frequency during quiet stance is increased with aging and more so in elderly people who fall.[60] Sway is also increased in many neurological diseases that affect sensory and/or motor systems and increased sway particularly in the mediolateral direction is associated with falls in a number of conditions, including PD.[61]

Postural sway can be abnormal in persons with PD, long before clinically evident balance impairment and prior to taking levodopa as shown in Fig. 3.5[62] and even in the prodromal phase.[63] Smoothness (jerkiness) of

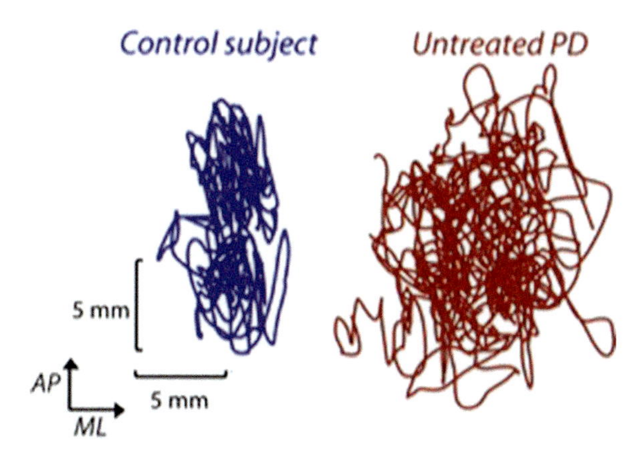

FIGURE 3.5 Representative postural sway, represented with the CoP excursion, in a healthy control and untreated PD. *AP*, anteroposterior CoP excursion; *ML*, mediolateral CoP excursion.

sway has been recently identified to be a sensitive discriminant measure in PD, and reduced smoothness may be due to rigidity. Later in the disease, postural sway in PD tends to be of higher amplitude, higher velocity, and lower smoothness than in normal controls.[64,65]

Sway amplitude has been related to the effectiveness of, or the stability achieved by, the postural control system, whereas mean velocity has been related to the amount of regulatory activity associated with this level of stability.[58,66] Thus, patients with untreated PD, who do not show clinical signs of balance or gait problems, may achieve the same level of stability as age-matched healthy control subjects, as reflected by normal sway area, but with more frequent corrections of postural sway, as reflected by abnormal higher derivatives of sway (increased sway velocity and frequency and reduced sway smoothness).

Disease severity can explain part of the controversies found in static posturography in patients with PD. In fact, several studies reported that the body sway of PD patients is similar to normal under quiet stance,[67,68] at least at the earlier stages of the disease,[69] whereas only more recent studies reported impairments of sway early in the disease. These controversies may also depend upon instructions, different sensory conditions, distractors (dual task), medications, and on how CoP displacement is quantified (sometime only the sagittal direction has been considered.[69] A more consistent result has been the increased sway in the mediolateral direction in PD.[70] In addition, the mediolateral measures were also associated with a history of falls and a poor balance performance.[60,71]

Standing balance is commonly thought to be an automatic process, largely independent of cortical control. However, performing another cognitive or motor task while standing, so called "dual-tasking," often increases postural sway area and velocity, especially in patients with PD.[72–74] Other papers demonstrate larger differences in sway while dual tasking in people with a history of falls compared to nonfallers.[73,75] Thus, higher cortical function is important in maintaining balance even when standing quietly.

E. Does PD affect sensory integration for balance?

Altering sensory conditions (such as closing the eyes or standing on foam pads) while standing has been used to identify subtle postural impairments. The ability to maintain a stable upright stance depends upon a complex integration of somatosensory, vestibular, visual stimuli with motor, premotor, and brainstem systems. In a healthy person, the somatosensory system contributes 70% to postural sway, whereas the vestibular and visual systems contribute 20% and 10%, respectively.[59,76]

An individual must reweight this sensory dependence for different environmental conditions in order to maintain their balance. For example, when standing on an unstable surface with eyes closed, healthy individuals' vestibular systems can contribute 100% to control postural sway.

Although they generally have no problem standing with eyes closed, individuals with severe PD often have difficulty standing on an unstable surface.[69] However, this does not necessarily mean that they have difficulty using vestibular information for balance. In fact, peripheral vestibular function is thought to be normal in subjects with PD.[77–79] A recent study suggests that subjects with PD rely more than normal on vestibular information to control postural sway in stance, irrespective of treatment with medication or DBS of the subthalamic nucleus.[80]

Standing on a compliant surface requires considerable proprioceptive-motor coordination that may be impaired in subjects with PD. There is increasing evidence that basal ganglia–related diseases, such as PD, are associated with kinesthetic deficits, including reduced tactile discrimination, poor joint kinesthesia, and overestimating of reaching and stepping when vision is not available.[81,82] There is also evidence that dopaminergic medication further depresses use of proprioception.[83,84] In addition, people with PD have difficulty recognizing small changes in surface orientation, which suggests reduced proprioception.[67]

Specific studies have developed methods that can accurately quantify the sources of sensory orientation information contributing to postural control.[59,85] These methods quantify a subject's relative reliance on visual, vestibular, and proprioceptive information for postural orientation in response to sensory stimuli. Young, healthy subjects typically rely heavily (70%) on proprioceptive cues, but shift toward decreased reliance on proprioception and increased reliance on vestibular and/or visual cues when stance is perturbed by rotation of the support surface. Similarly, when perturbed by rotation of a visual surround, subjects decrease their reliance on visual orientation cues.

Interestingly, on a sample of eight subjects with early to moderate PD tested with this paradigm, subjects with PD can reweight proprioceptive, visual, and vestibular information in response to changing sensory stimuli.[85a] Each subject with PD and age-matched control decreased reliance on proprioceptive information as the surface-stimulus amplitude increased from 1 to 6 degrees (Fig. 3.6). Although patients with PD were able to reweight away from surface- and visual stimuli when these sensory inputs were unreliably related to postural sway, the change in proprioceptive weights was smaller in subjects with PD than control subjects for small changes in surface angle amplitudes. This impairment in sensory reweighting between two similar stimulus amplitudes may be due to increased noise in the basal ganglia circuits. Previous studies have

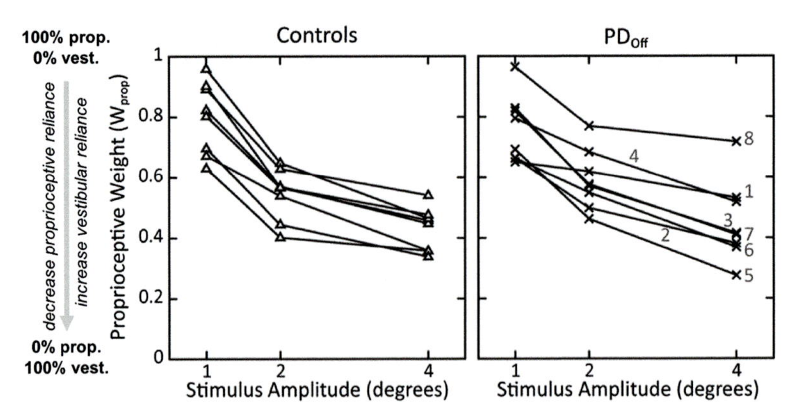

FIGURE 3.6 Sensory reweighting in individual subjects, healthy controls on the left and people with PD on the right. *Adapted from Feller K, Peterka RJ, Horak FB. Sensory Re-weighting for Postural Control in Parkinson's Disease. Front. Hum. Neurosci. 2019.*

suggested that there is an increase in noise in dopaminergic circuitry as dopamine transmitter decreases with PD.[86,87] This increase in noise in the basal ganglia circuits could be responsible for abnormal sensory integration for postural control in PD. For example, increased noise could increase the amount of time to switch between sensory conditions, translating to an inability to rapidly change sensory weighting for different situations.[88]

F. Does levodopa improve standing balance?

Results in camptocormia or Pisa syndrome after levodopa intake are scarce. In summary, levodopa therapy aggravated the camptocormia in three patients, with slight improvement in the other five in a small study of eight subjects with PD.[89] Holler and colleagues[90] described two patients with PD and camptocormia in whom dopaminergic treatment had no effect on posture. Although levodopa replacement therapy does not seem to improve Pisa syndrome or camptocormia, imaging with iodine-123-beta-carbomethoxy-3 beta-(4-iodophenyltropane) (^{123}I-β-CIT) single photon emission computed tomography (SPECT) revealed reduced striatal dopamine binding in patients with Pisa syndrome.[90] The camptocormia improved with sensory tricks, supporting the theory of camptocormia as a type of dystonia.

Levodopa improves limits of stability. Levodopa intake increased the limits of stability in subjects with PD, but did not change postural strategies used to reach such limits (Fig. 3.3). It is possible that reduced rigidity played a role in allowing larger stability limits with levodopa, although no significant correlations were found between limits of stability and the UPDRS measures of rigidity. Indeed, previous studies showed that PD subjects' background EMG is quieter and CoM moves farther and faster in response to external perturbations and during quiet stance when On levodopa, consistent with reduced rigidity.[91] Increased functional stability limits in the On state may be related to reduction of leg, and not axial, rigidity, because a previous study showed no reduction of axial rigidity with levodopa.[52]

Levodopa does not alter axial postural tone. A consistent body of literature finds dopaminergic medications to be ineffective for axial hypertonia.[1,92,93] Recent studies that directly measure axial tone in PD subjects[52,53] also showed that hypertonicity in the neck, trunk, and hips is insensitive to levodopa. This agrees with other previous findings that high amplitudes of muscle activity during stance are significantly reduced in the distal, but not proximal, muscles with levodopa.[1,94] Taken together, these findings support the theory that axial musculature is controlled by nondopaminergic central structures and pathways that are separate from the levodopa-sensitive pathways for distal muscles.[95]

Levodopa appears to help postural sway early in the disease but worsens it later in the disease (Fig. 3.7). A small study by Beuter recently showed that postural sway improved in 10 subjects with mild PD (diagnosis within 6 years) when On levodopa, compared to the Off condition. The preliminary results suggest that postural control mechanisms are affected early in PD and may be modulated by dopamine.[96]

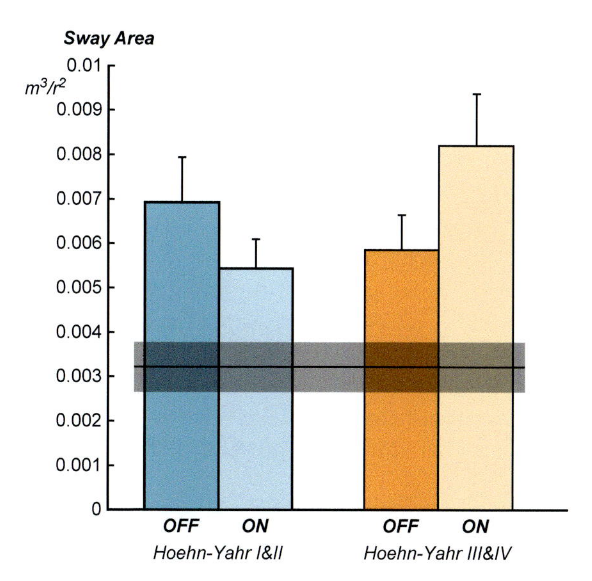

FIGURE 3.7 Mean and standard deviation (SD) of sway area during 30 degrees of quiet standing in mild PD, H&Y I-II and moderate PD, H&Y III & IV. Healthy controls are represented in the gray shaded area (unpublished data).

However, the picture gets more complicated when looking at effects of levodopa in moderate to severe stages of PD.[97,98] Particularly, treatment with levodopa increases postural sway in patients with advanced PD. Some of the increase in sway may be attributed to levodopa-induced dyskinesia.[97,99] Levodopa-induced dyskinesia may, in fact, be part of the reason that PD patients fall when "On." The interventions often helpful in improving other PD symptoms, dopaminergic medications and deep brain stimulation, do not consistently improve balance and falls.[100,101] Degeneration in other brain areas, possibly the adrenergic locus coeruleus or the cholinergic/glutaminergic pedunculopontine nucleus or frontal cortex may be the origin of some of the balance problems associated with PD.[102]

G. Do other medications influence balance?

It has been speculated that decline in the neurotransmitter, acetylcholine (Ach), may be related to impaired postural sway and falls in PD.[103] PET studies of Ach show postural sway correlated significantly with the amount of Ach in the thalamus in subjects with PD.[103] It is possible that cholinesterase inhibitors could help reduce falls through direct effects on balance, improvements in cognition, or more complex integrating systems. Previous studies indicated that integration of sensory information is affected in PD,[69,104] and that patients with PD show excessive postural sway in conditions with limited or inappropriate sensory feedback.[68,105] Therefore, cholinergic treatment in PD may ameliorate postural instability particularly in conditions in which somatosensory information is altered. In fact, two studies found that the cholinesterase inhibitor, donepezil, reduced mediolateral postural sway and falls specifically in challenging sensory conditions with unstable support surface,[106,107] however these results would need to be reproduced in a larger cohort.

Antidepressants alone, or in combination with benzodiazepines, were found to be associated with an increased frequency of falls, nearly two times higher than a similar group of patients with PD not taking psychotropic medication.[108–110] Antidepressants have several mechanisms of action, such as inhibiting serotonin reuptake or serotonin—norepinephrine reuptake, acting as serotonin modulators, or inhibiting dopamine and norepinephrine reuptake. However, the mechanisms by which these medications can affect posture and locomotion leading to falls remain unknown.

Benzodiazepines cause acute adverse effects such as drowsiness, motor incoordination, or increased reaction time.[111] The sedative properties of the benzodiazepines could possibly exacerbate gait issues in patients with PD. However, neuropsychiatric symptoms need to be treated aggressively to improve patients' and caregivers' quality of life. It is hard to know when the associated increased risk of falling is worth the price of antidepressant or benzodiazepine efficacy on nonmotor symptoms. There is definitely a need to better understand how these medications affect patients with PD, and more importantly, to learn how subclasses of these drugs or even different dosages affect PD.

H. What are the effects of deep brain stimulation on standing posture?

In patients with advanced PD, deep brain stimulation (DBS) of the subthalamic nucleus (STN) has been proven to improve motor symptoms and activities of daily living, as well as to reduce dyskinesias, providing a significant and persistent antiparkinsonian effect.[112] It is generally accepted that DBS ameliorates the symptoms that respond to antiparkinson medications but has no effect on the symptoms that do not, such as freezing and falls. However, a few reports,[23,113–115] suggest that the camptocormia, bent knees, and tiptoeing responded favorably to DBS stimulation, just like the other levodopa-responsive symptoms. But these cases are exceptions rather than the rule. These findings may suggest that in these cases the camptocormia was attributable to a central mechanism, such as dystonia.

Most of the existing studies investigating the effects of DBS on postural deformity are related to camptocormia, whereas a few have addressed Pisa syndrome.[12] Preliminary data suggest that Pisa syndrome in patients with PD may be ameliorated by STN DBS.[115] The clinical course of 18 patients with significant postural abnormality (score $\geq$ 2 on item 28 of the UPDRS III) that underwent bilateral STN DBS was reviewed. A total of 8 patients suffered from camptocormia and the remaining 10 patients were considered to have Pisa. Most patients had significant motor fluctuations as a result of levodopa therapy. Of the 13 patients with moderate postural abnormality (score of 2 on item 28 UPDRS III), 9 improved soon after surgery. One patient relapsed, and 2 patients improved gradually over time, whereas 2 patients showed no improvement. Of the 5 patients with severe postural abnormality (score of 3 or 4 on item 28 of the UPDRS III), 2 improved slightly in the long-term follow-up period, whereas the remaining patients did not improve at all.

Recently, Shih and colleagues[116] described progressive improvement of Pisa syndrome after left PPN DBS in a case study of a 62-year-old woman with right Pisa syndrome and postural instability. Ricciardi and colleagues[117] showed that unilateral PPN stimulation provided short-term benefits in a patient with Pisa. At postsurgical assessment, except for improvement in Pisa, no other relevant changes in the UPDRS III scores were noted. However, during the following years, the patient's posture progressively worsened, as did his motor and cognitive functions.

Using posturography, a study[118] showed that STN DBS alone reversed medication—induced postural instability, reducing sway velocity. To the best of our knowledge, the study by Nantel is the first comparing objective measures of postural instability before (both OFF and ON medication) and after STN DBS (both OFF and ON medication and DBS). In fact, previous studies[66,119] reported changes in postural sway only after GPi and STN DBS, showing that the combination of DBS and levodopa resulted in improvements in sway compared to levodopa alone (DBS off).

More recently, our group examined postural sway before and after (6 months) in a group of 33 subjects with PD randomized to either GPi or STN DBS. Results showed a marked improvement in postural sway, particularly in sway velocity, after DBS in the GPi but not in the STN groups (Fig. 3.8). Specifically, sway velocity significantly decreased in the

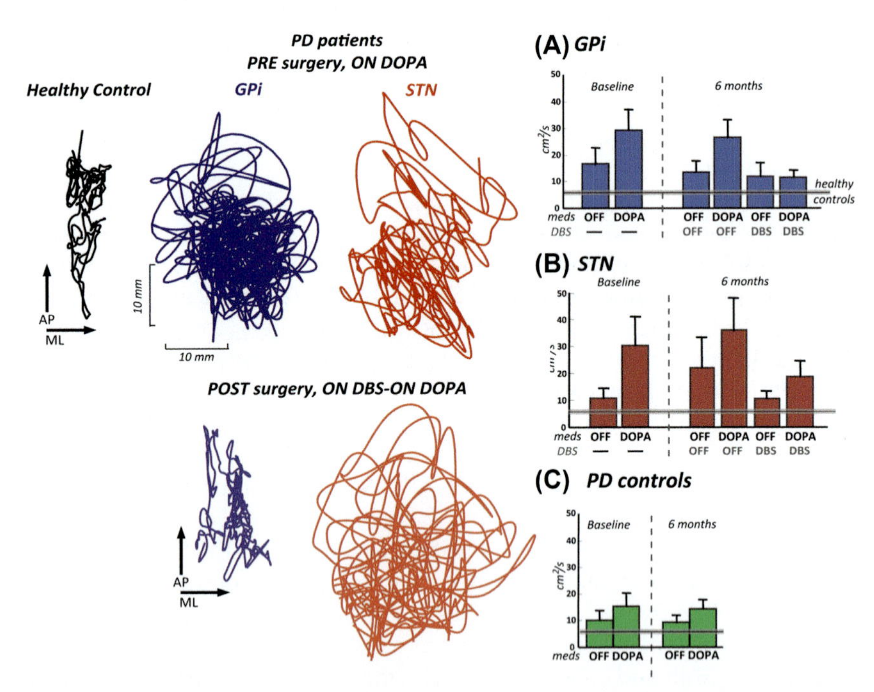

FIGURE 3.8 *Left panel*: Representative CoP traces for 60s of quiet standing in a healthy control and in 2 subject with PD (before and after GPi and STN DBS). *Right panel*: Mean and SD of PD group before and 6 months after GPi (A) and STN (B) DBS. The GPi group showed no change in postural sway after DBS, which was larger than controls, especially ON Levodopa. The STN group should increased postural sway in the OFF state after DBS surgery. An additional control group of PD subjects (C) tested 6 months apart OFF and ON levodopa showed no change in postural sway over time.

best state after GPi DBS compared to before surgery; instead, there was no significant change in the STN DBS group.

I. Can rehabilitation improve standing posture?

Several small studies have investigated the effects of rehabilitation on postural alignment, limits of stability, and postural sway with promising results. Rehabilitation may improve postural alignment but the best intervention and for how long the program should last is uncertain.

For example, two randomized controlled studies,[120,121] one open-label study[122] and two case reports,[123,124] have provided evidence that motor rehabilitation is an effective tool for PD patients with Pisa Syndrome. Although rehabilitation programs reported are heterogeneous, they share the concept of high intensity with 4–5 days a week for 2 or 4 weeks. Moreover, in two studies rehabilitation was combined to Botox[120,123] and in one to Kinesio Taping.[121] In the RCT by Tassorelli et al., the effect on kinematic trunk measures were more pronounced for both lateral and anterior bending when rehabilitation was combined with Botox injections (injections in the more hyperactive muscles of the trunk). In another study, the addition of Kinesio Taping to rehabilitation did not further improve the positive outcome produced by rehabilitation alone. A review for the management of Pisa syndrome in PD pointed out that rehabilitation programs should be preferred to surgical approaches that may lead to secondary complications.[12] Stretching exercises were included in all of the mentioned studies, whereas proprioceptive and tactile stimulation, postural reeducation through active movement execution,[121] or strengthening exercises and concomitant balance and gait training[122] were included in a few studies. The duration of the benefit of the rehabilitation program was assessed during a maximal period of 6 months, with most of the outcome measures improving postural alignment up to 3 months with a waning at 6 months.

Limits of stability can be improved in patients with PD after four weeks of repetitive step training using preparatory cues. Specifically, significant changes were found in reaction time, movement velocity, and endpoint excursion of voluntarily leaning to forward and backward limits.[125] However, there are also negative findings on the effect of step training in PD for limits of stability.[126,127] Variations in the findings could not be related to measurement outcome, since all studies used the Neurocom Balance Master to measure the limits of stability parameters. One study reported an improvement in LOS but failed to see a between-group difference for all LOS variables,[127] and another study demonstrated that the exercise group had significantly more improvements in limits of stability than control subjects.[125] However, the subjects in the first study received only voluntary step training, while the subjects in the second study received precued, compensatory stepping as well as voluntary step training which may be more effective in enhancing limits of stability. Postural strategies of LOS have been found to comprise a postural preparatory and an executive phase. Further studies are needed to understand the mechanisms behind improvements in limits of stability, and the involvement of the limits of stability preparatory phase on the limits of stability amplitude.

Only one study evaluated the effects of rehabilitation on postural sway.[128] Although postural sway is considered a sensitive measure of balance control and increased sway is associated with falls, it is not a measure typically assessed in the clinic. In the study from King et al., 44 subjects with PD completed a randomized training of 4 weeks in a treadmill group or Agility Boot Camp group. Mediolateral sway amplitude significantly decreased in the Agility Boot Camp group but not in the treadmill group. In contrast, several clinical measures such as the Berg Balance Scale and the UPDRS did not show differences between treatment groups. These findings suggest that an important reason for a lack of rehabilitation program effectiveness may be due to insensitive, clinical outcome measures.

Several studies have shown that light touch (<100 g) of a single finger on a stable support while standing can reduce postural sway even more than adding vision.[129] Subjects with PD also reduce postural sway with light touch, and this reduction is associated with increased axial tone.[130] Thus, use of a cane, walker, or touch of a stable support may improve balance and reduce falls in patients with PD via increasing axial tone.

Highlights

- Flexed postural alignment such as stooped posture, camptocormia, and Pisa syndrome impairs balance.
- Limits of stability are impaired in PD, especially in the forward direction.
- Axial tone is higher than normal in PD, and neck tone is related to functional mobility.
- Postural sway is impaired, even in untreated PD.
- People with PD may have difficulty quickly integrating proprioception for standing balance.
- Although levodopa improves limits of stability, it does not improve flexed alignment or axial tone, and even worsens postural sway later in the disease.

- DBS may improve postural alignment, but findings are not yet conclusive. GPi, but not STN DBS, seems to improve postural sway.
- Rehabilitation has been found beneficial for postural alignment, limits of stability, axial tone, and postural sway, although the best type of rehabilitation is not clear yet.

References

1. Burleigh A, Horak F, Nutt J, Frank J. Levodopa reduces muscle tone and lower extremity tremor in Parkinson's disease. *The Canadian Journal of Neurological Sciences* 1995; **22**(4):280–5.
2. Bloem BR, Beckley DJ, van Dijk JG. Are automatic postural responses in patients with Parkinson's disease abnormal due to their stooped posture? *Experimental Brain Research* 1999;**124**(4):481–8.
3. Jacobs JV, Dimitrova DM, Nutt JG, Horak FB. Can stooped posture explain multidirectional postural instability in patients with Parkinson's disease? *Experimental Brain Research* 2005;**166**(1):78–88.
4. Horak FB, Macpherson JM. Postural orientation and equilibrium. *Comprehensive Physiology* 2011:255–92.
5. Bronstein AM, Pavlou M. Balance. *Handbook of Clinical Neurology* 2013;**110**:189–208.
6. Vaugoyeau M, Viel S, Assaiante C, Amblard B, Azulay JP. Impaired vertical postural control and proprioceptive integration deficits in Parkinson's disease. *Neuroscience* 2007;**146**(2):852–63.
7. Carpenter MG, Bloem BR. Postural control in Parkinson patients: a proprioceptive problem? *Experimental Neurology* 2011;**227**(1):26–30.
8. Ashour R, Jankovic J. Joint and skeletal deformities in Parkinson's disease, multiple system atrophy, and progressive supranuclear palsy. *Movement Disorders: Official Journal of the Movement Disorder Society* 2006;**21**(11):1856–63.
9. Benatru I, Vaugoyeau M, Azulay JP. Postural disorders in Parkinson's disease. *Neurophysiologie Clinique = Clinical Neurophysiology* 2008;**38**(6):459–65.
10. Doherty KM, van de Warrenburg BP, Peralta MC, et al. Postural deformities in Parkinson's disease. *The Lancet Neurology* 2011;**10**(6):538–49.
11. Tassorelli C, Furnari A, Buscone S, et al. Pisa syndrome in Parkinson's disease: clinical, electromyographic, and radiological characterization. *Movement Disorders: Official Journal of the Movement Disorder Society* 2012;**27**(2):227–35.
12. Tinazzi M, Geroin C, Gandolfi M, et al. Pisa syndrome in Parkinson's disease: an integrated approach from pathophysiology to management. *Movement Disorders: Official Journal of the Movement Disorder Society* 2016;**31**(12):1785–95.
13. Gurfinkel V, Cacciatore TW, Cordo P, Horak F, Nutt J, Skoss R. Postural muscle tone in the body axis of healthy humans. *Journal of Neurophysiology* 2006;**96**(5):2678–87.
14. St George RJ, Gurfinkel VS, Kraakevik J, Nutt JG, Horak FB. Case studies in neuroscience: a dissociation of balance and posture demonstrated by camptocormia. *Journal of Neurophysiology* 2017. jn.00582.2017.
15. Margraf NG, Wrede A, Rohr A, et al. Camptocormia in idiopathic Parkinson's disease: a focal myopathy of the paravertebral muscles. *Movement Disorders: Official Journal of the Movement Disorder Society* 2010;**25**(5):542–51.
16. Spuler S, Krug H, Klein C, et al. Myopathy causing camptocormia in idiopathic Parkinson's disease: a multidisciplinary approach. *Movement Disorders: Official Journal of the Movement Disorder Society* 2010;**25**(5):552–9.

17. Laroche M, Ricq G, Delisle MB, Campech M, Marque P. Bent spine syndrome: computed tomographic study and isokinetic evaluation. *Muscle and Nerve* 2002;**25**(2): 189–93.
18. Mahjneh I, Marconi G, Paetau A, Saarinen A, Salmi T, Somer H. Axial myopathy—an unrecognised entity. *Journal of Neurology* 2002;**249**(6):730–4.
19. Laroche M, Delisle MB, Aziza R, Lagarrigue J, Mazieres B. Is camptocormia a primary muscular disease? *Spine* 1995;**20**(9):1011–6.
20. Lepoutre AC, Devos D, Blanchard-Dauphin A, et al. A specific clinical pattern of camptocormia in Parkinson's disease. *Journal of Neurology Neurosurgery and Psychiatry* 2006; **77**(11):1229–34.
21. Melamed E, Djaldetti R. Camptocormia in Parkinson's disease. *Journal of Neurology* 2006;**253**(Suppl. 7). Vii14-V16.
22. Bloch F, Houeto JL, Tezenas du Montcel S, et al. Parkinson's disease with camptocormia. *Journal of Neurology Neurosurgery and Psychiatry* 2006;**77**(11):1223–8.
23. Azher SN, Jankovic J. Camptocormia: pathogenesis, classification, and response to therapy. *Neurology* 2005;**65**(3):355–9.
24. Barone P, Santangelo G, Amboni M, Pellecchia MT, Vitale C. Pisa syndrome in Parkinson's disease and parkinsonism: clinical features, pathophysiology, and treatment. *The Lancet Neurology* 2016;**15**(10):1063–74.
25. Galati S, Moller JC, Stadler C. Ropinirole-induced Pisa syndrome in Parkinson disease. *Clinical Neuropharmacology* 2014;**37**(2):58–9.
26. Michel SF, Arias Carrion O, Correa TE, Alejandro PL, Micheli F. Pisa syndrome. *Clinical Neuropharmacology* 2015;**38**(4):135–40.
27. Suzuki T, Matsuzaka H. Drug-induced Pisa syndrome (pleurothotonus): epidemiology and management. *CNS Drugs* 2002;**16**(3):165–74.
28. Castrioto A, Piscicelli C, Perennou D, Krack P, Debu B. The pathogenesis of Pisa syndrome in Parkinson's disease. *Movement Disorders: Official Journal of the Movement Disorder Society* 2014;**29**(9):1100–7.
29. Cannas A, Solla P, Floris G, et al. Reversible Pisa syndrome in patients with Parkinson's disease on dopaminergic therapy. *Journal of Neurology* 2009;**256**(3):390–5.
30. Fasano A, Di Matteo A, Vitale C, et al. Reversible Pisa syndrome in patients with Parkinson's disease on rasagiline therapy. *Movement Disorders: Official Journal of the Movement Disorder Society* 2011;**26**(14):2578–80.
31. Ungerstedt U, Butcher LL, Butcher SG, Anden NE, Fuxe K. Direct chemical stimulation of dopaminergic mechanisms in the neostriatum of the rat. *Brain Research* 1969;**14**(2): 461–71.
32. Boonstra TA, van der Kooij H, Munneke M, Bloem BR. Gait disorders and balance disturbances in Parkinson's disease: clinical update and pathophysiology. *Current Opinion in Neurology* 2008;**21**(4):461–71.
33. Vitale C, Marcelli V, Furia T, et al. Vestibular impairment and adaptive postural imbalance in parkinsonian patients with lateral trunk flexion. *Movement Disorders: Official Journal of the Movement Disorder Society* 2011;**26**(8):1458–63.
34. Schieppati M, Hugon M, Grasso M, Nardone A, Galante M. The limits of equilibrium in young and elderly normal subjects and in parkinsonians. *Electroencephalography and Clinical Neurophysiology* 1994;**93**(4):286–98.
35. Mancini M, Rocchi L, Horak FB, Chiari L. Effects of Parkinson's disease and levodopa on functional limits of stability. *Clinical Biomechanics* 2008;**23**(4):450–8.
36. Horak FB, Dimitrova D, Nutt JG. Direction-specific postural instability in subjects with Parkinson's disease. *Experimental Neurology* 2005;**193**(2):504–21.
37. Burleigh-Jacobs A, Horak FB, Nutt JG, Obeso JA. Step initiation in Parkinson's disease: influence of levodopa and external sensory triggers. *Movement Disorders: Official Journal of the Movement Disorder Society* 1997;**12**(2):206–15.

38. Ferrarin M, Lopiano L, Rizzone M, et al. Quantitative analysis of gait in Parkinson's disease: a pilot study on the effects of bilateral sub-thalamic stimulation. *Gait and Posture* 2002;**16**(2):135–48.
39. Rocchi L, Chiari L, Mancini M, Carlson-Kuhta P, Gross A, Horak FB. Step initiation in Parkinson's disease: influence of initial stance conditions. *Neuroscience Letters* 2006; **406**(1–2):128–32.
40. Mak MK, Cole JH. Movement dysfunction in patients with Parkinson's disease: a literature review. *Australian Journal of Physiotherapy* 1991;**37**(1):7–17.
41. Horak FB, Henry SM, Shumway-Cook A. Postural perturbations: new insights for treatment of balance disorders. *Physical Therapy* 1997;**77**(5):517–33.
42. Chong RK, Horak FB, Woollacott MH. Parkinson's disease impairs the ability to change set quickly. *Journal of the Neurological Sciences* 2000;**175**(1):57–70.
43. Stack EL, Ashburn AM. Impaired bed mobility and disordered sleep in Parkinson's disease. *Movement Disorders: Official Journal of the Movement Disorder Society* 2006; **21**(9):1340–2.
44. Nutt J, Hammerstad JP, Gancher ST. *Parkinson's disease*. St. Louis, MO: Mosy Year Book; 1992.
45. Crenna P, Carpinella I, Rabuffetti M, et al. The association between impaired turning and normal straight walking in Parkinson's disease. *Gait and Posture* 2007;**26**(2):172–8.
46. Mesure S, Azulay JP, Pouget J, Amblard B. Strategies of segmental stabilization during gait in Parkinson's disease. *Experimental Brain Research* 1999;**129**(4):573–81.
47. Vaugoyeau M, Viallet F, Aurenty R, Assaiante C, Mesure S, Massion J. Axial rotation in Parkinson's disease. *Journal of Neurology Neurosurgery and Psychiatry* 2006;**77**(7):815–21.
48. Nagumo K, Hirayama K. A study on truncal rigidity in parkinsonism–evaluation of diagnostic test and electrophysiological study. *Rinsho shinkeigaku = Clinical Neurology* 1993;**33**(1):27–35.
49. Nagumo K, Hirayama K. Axial (neck and trunk) rigidity in Parkinson's disease, striatonigral degeneration and progressive supranuclear palsy. *Rinsho shinkeigaku = Clinical neurology* 1996;**36**(10):1129–35.
50. Mak MK, Wong EC, Hui-Chan CW. Quantitative measurement of trunk rigidity in parkinsonian patients. *Journal of Neurology* 2007;**254**(2):202–9.
51. Sherrington CS. A mammalian spinal preparation. *The Journal of Physiology* 1909;**38**(5): 375–83.
52. Wright WG, Gurfinkel VS, Nutt J, Horak FB, Cordo PJ. Axial hypertonicity in Parkinson's disease: direct measurements of trunk and hip torque. *Experimental Neurology* 2007;**208**(1):38–46.
53. Franzen E, Paquette C, Gurfinkel VS, Cordo PJ, Nutt JG, Horak FB. Reduced performance in balance, walking and turning tasks is associated with increased neck tone in Parkinson's disease. *Experimental Neurology* 2009;**219**(2):430–8.
54. Stack EL, Ashburn AM, Jupp KE. Strategies used by people with Parkinson's disease who report difficulty turning. *Parkinsonism and Related Disorders* 2006;**12**(2):87–92.
55. Horak FB, Nashner LM. Central programming of postural movements: adaptation to altered support-surface configurations. *Journal of Neurophysiology* 1986;**55**(6):1369–81.
56. Henry SM, Fung J, Horak FB. Control of stance during lateral and anterior/posterior surface translations. *IEEE Transactions on Rehabilitation Engineering: A Publication of the IEEE Engineering in Medicine and Biology Society* 1998;**6**(1):32–42.
57. Dimitrova D, Nutt J, Horak FB. Abnormal force patterns for multidirectional postural responses in patients with Parkinson's disease. *Experimental Brain Research* 2004;**156**(2): 183–95.
58. Maurer C, Mergner T, Peterka RJ. Multisensory control of human upright stance. *Experimental Brain Research* 2006;**171**(2):231–50.

59. Peterka RJ. Sensorimotor integration in human postural control. *Journal of Neurophysiology* 2002;**88**(3):1097–118.
60. Piirtola M, Era P. Force platform measurements as predictors of falls among older people - a review. *Gerontology* 2006;**52**(1):1–16.
61. Schoneburg B, Mancini M, Horak F, Nutt JG. Framework for understanding balance dysfunction in Parkinson's disease. *Movement Disorders: Official Journal of the Movement Disorder Society* 2013;**28**(11):1474–82.
62. Mancini M, Horak FB, Zampieri C, Carlson-Kuhta P, Nutt JG, Chiari L. Trunk accelerometry reveals postural instability in untreated Parkinson's disease. *Parkinsonism and Related Disorders* 2011;**17**(7):557–62.
63. Maetzler W, Hausdorff JM. Motor signs in the prodromal phase of Parkinson's disease. *Movement Disorders: Official Journal of the Movement Disorder Society* 2012;**27**(5):627–33.
64. Mancini M, Zampieri C, Carlson-Kuhta P, Chiari L, Horak FB. Anticipatory postural adjustments prior to step initiation are hypometric in untreated Parkinson's disease: an accelerometer-based approach. *European Journal of Neurology* 2009;**16**(9):1028–34.
65. Mancini M, Carlson-Kuhta P, Zampieri C, Nutt JG, Chiari L, Horak FB. Postural sway as a marker of progression in Parkinson's disease: a pilot longitudinal study. *Gait and Posture* 2012;**36**(3):471–6.
66. Rocchi L, Chiari L, Horak FB. Effects of deep brain stimulation and levodopa on postural sway in Parkinson's disease. *Journal of Neurology Neurosurgery and Psychiatry* 2002;**73**(3):267–74.
67. Bronstein AM, Hood JD, Gresty MA, Panagi C. Visual control of balance in cerebellar and parkinsonian syndromes. *Brain: A Journal of Neurology* 1990;**113**(Pt 3):767–79.
68. Bronte-Stewart HM, Minn AY, Rodrigues K, Buckley EL, Nashner LM. Postural instability in idiopathic Parkinson's disease: the role of medication and unilateral pallidotomy. *Brain* 2002;**125**(Pt 9):2100–14.
69. Frenklach A, Louie S, Koop MM, Bronte-Stewart H. Excessive postural sway and the risk of falls at different stages of Parkinson's disease. *Movement Disorders: Official Journal of the Movement Disorder Society* 2009;**24**(3):377–85.
70. Mitchell SL, Collins JJ, De Luca CJ, Burrows A, Lipsitz LA. Open-loop and closed-loop postural control mechanisms in Parkinson's disease: increased mediolateral activity during quiet standing. *Neuroscience Letters* 1995;**197**(2):133–6.
71. Visser JE, Carpenter MG, van der Kooij H, Bloem BR. The clinical utility of posturography. *Clinical Neurophysiology: Official Journal of the International Federation of Clinical Neurophysiology* 2008;**119**(11):2424–36.
72. Ashburn A, Stack E, Pickering RM, Ward CD. A community-dwelling sample of people with Parkinson's disease: characteristics of fallers and non-fallers. *Age and Ageing* 2001; **30**(1):47–52.
73. Marchese R, Bove M, Abbruzzese G. Effect of cognitive and motor tasks on postural stability in Parkinson's disease: a posturographic study. *Movement Disorders: Official Journal of the Movement Disorder Society* 2003;**18**(6):652–8.
74. Morris M, Iansek R, Smithson F, Huxham F. Postural instability in Parkinson's disease: a comparison with and without a concurrent task. *Gait and Posture* 2000;**12**(3):205–16.
75. Bekkers EMJ, Dockx K, Devan S, et al. The impact of dual-tasking on postural stability in people with Parkinson's disease with and without freezing of gait. *Neurorehabilitation and Neural Repair* 2018;**32**(2):166–74.
76. Peterka RJ, Benolken MS. Role of somatosensory and vestibular cues in attenuating visually induced human postural sway. *Experimental Brain Research* 1995;**105**(1):101–10.
77. Bronstein AM, Yardley L, Moore AP, Cleeves L. Visually and posturally mediated tilt illusion in Parkinson's disease and in labyrinthine defective subjects. *Neurology* 1996; **47**(3):651–6.
78. Pastor MA, Day BL, Marsden CD. Vestibular induced postural responses in Parkinson's disease. *Brain: A Journal of Neurology* 1993;**116**(Pt 5):1177–90.

79. Chong RK, Horak FB, Frank J, Kaye J. Sensory organization for balance: specific deficits in Alzheimer's but not in Parkinson's disease. *The Journals of Gerontology Series A, Biological Sciences and Medical Sciences* 1999;**54**(3):M122–8.
80. Maurer C, Mergner T, Xie J, Faist M, Pollak P, Lucking CH. Effect of chronic bilateral subthalamic nucleus (STN) stimulation on postural control in Parkinson's disease. *Brain: A Journal of Neurology* 2003;**126**(Pt 5):1146–63.
81. Tagliabue M, Ferrigno G, Horak F. Effects of Parkinson's disease on proprioceptive control of posture and reaching while standing. *Neuroscience* 2009;**158**(4):1206–14.
82. Jacobs JV, Fujiwara K, Tomita H, Furune N, Kunita K, Horak FB. Changes in the activity of the cerebral cortex relate to postural response modification when warned of a perturbation. *Clinical Neurophysiology: Official Journal of the International Federation of Clinical Neurophysiology* 2008;**119**(6):1431–42.
83. Mongeon D, Blanchet P, Messier J. Impact of Parkinson's disease and dopaminergic medication on proprioceptive processing. *Neuroscience* 2009;**158**(2):426–40.
84. O'Suilleabhain P, Bullard J, Dewey RB. Proprioception in Parkinson's disease is acutely depressed by dopaminergic medications. *Journal of Neurology Neurosurgery and Psychiatry* 2001;**71**(5):607–10.
85. Peterka RJ, Murchison CF, Parrington L, Fino PC, King LA. Implementation of a central sensorimotor integration test for characterization of human balance control during stance. *Frontiers in Neurology* 2018;**9**:1045.
85a. Feller KJ, Peterka RJ, Horak FB. Sensory re-weighting for postural control in Parkinson's disease. *Front Hum Neurosci* 2019;**13**:126. https://doi.org/10.3389/fnhum.2019.00126. eCollection 2019.
86. Obeso JA, Rodriguez-Oroz MC, Rodriguez M, et al. Pathophysiologic basis of surgery for Parkinson's disease. *Neurology* 2000;**55**(12 Suppl. 6):S7–12.
87. Okun MS, Vitek JL. Lesion therapy for Parkinson's disease and other movement disorders: update and controversies. *Movement Disorders: Official Journal of the Movement Disorder Society* 2004;**19**(4):375–89.
88. Peterka RJ, Benolken MS. Relation between perception of vertical axis rotation and vestibulo-ocular reflex symmetry. *Journal of Vestibular Research: Equilibrium and Orientation* 1992;**2**(1):59–69.
89. Djaldetti R, Mosberg-Galili R, Sroka H, Merims D, Melamed E. Camptocormia (bent spine) in patients with Parkinson's disease–characterization and possible pathogenesis of an unusual phenomenon. *Movement Disorders: Official Journal of the Movement Disorder Society* 1999;**14**(3):443–7.
90. Holler I, Dirnberger G, Pirker W, Auff E, Gerschlager W. Camptocormia in idiopathic Parkinson's disease: [(123)I]beta-CIT SPECT and clinical characteristics. *European Neurology* 2003;**50**(2):118–20.
91. Horak FB, Frank J, Nutt J. Effects of dopamine on postural control in parkinsonian subjects: scaling, set, and tone. *Journal of Neurophysiology* 1996;**75**(6):2380–96.
92. Pavese N, Evans AH, Tai YF, et al. Clinical correlates of levodopa-induced dopamine release in Parkinson disease: a PET study. *Neurology* 2006;**67**(9):1612–7.
93. Weinrich M, Koch K, Garcia F, Angel RW. Axial versus distal motor impairment in Parkinson's disease. *Neurology* 1988;**38**(4):540–5.
94. Bejjani BP, Gervais D, Arnulf I, et al. Axial parkinsonian symptoms can be improved: the role of levodopa and bilateral subthalamic stimulation. *Journal of Neurology Neurosurgery and Psychiatry* 2000;**68**(5):595–600.
95. Bloem BR. Postural instability in Parkinson's disease. *Clinical Neurology and Neurosurgery* 1992;**94**(Suppl. l):S41–5.
96. Beuter A, Hernandez R, Rigal R, Modolo J, Blanchet PJ. Postural sway and effect of levodopa in early Parkinson's disease. *The Canadian Journal of Neurological Sciences* 2008;**35**(1):65–8.

97. Curtze C, Nutt JG, Carlson-Kuhta P, Mancini M, Horak FB. Objective gait and balance impairments relate to balance confidence and perceived mobility in people with Parkinson disease. *Physical Therapy* 2016;**96**(11):1734–43.
98. Curtze C, Nutt JG, Carlson-Kuhta P, Mancini M, Horak FB. Levodopa is a double-edged sword for balance and gait in people with Parkinson's disease. *Movement Disorders: Official Journal of the Movement Disorder Society* 2015;**30**(10):1361–70.
99. Chung KA, Lobb BM, Nutt JG, McNames J, Horak F. Objective measurement of dyskinesia in Parkinson's disease using a force plate. *Movement Disorders: Official Journal of the Movement Disorder Society* 2010;**25**(5):602–8.
100. Weaver FM, Follett K, Stern M, et al. Bilateral deep brain stimulation vs best medical therapy for patients with advanced Parkinson disease: a randomized controlled trial. *JAMA* 2009;**301**(1):63–73.
101. Bloem BR, Beckley DJ, van Dijk JG, Zwinderman AH, Remler MP, Roos RA. Influence of dopaminergic medication on automatic postural responses and balance impairment in Parkinson's disease. *Movement Disorders: Official Journal of the Movement Disorder Society* 1996;**11**(5):509–21.
102. Bloem BR, Hausdorff JM, Visser JE, Giladi N. Falls and freezing of gait in Parkinson's disease: a review of two interconnected, episodic phenomena. *Movement Disorders* 2004;**19**(8):871–84.
103. Bohnen NI, Muller ML, Koeppe RA, et al. History of falls in Parkinson disease is associated with reduced cholinergic activity. *Neurology* 2009;**73**(20):1670–6.
104. Colnat-Coulbois S, Gauchard GC, Maillard L, et al. Management of postural sensory conflict and dynamic balance control in late-stage Parkinson's disease. *Neuroscience* 2011;**193**:363–9.
105. Horak FB, Nutt JG, Nashner LM. Postural inflexibility in parkinsonian subjects. *Journal of the Neurological Sciences* 1992;**111**(1):46–58.
106. Chung KA, Lobb BM, Nutt JG, Horak FB. Effects of a central cholinesterase inhibitor on reducing falls in Parkinson disease. *Neurology* 2010;**75**(14):1263–9.
107. Hiller ALP, Nutt JG, Mancini M, et al. Are cholinesterase inhibitors effective in improving balance in Parkinson's disease? *Journal of Neurological Disorders* 2015;**S2**(002).
108. Martinez-Ramirez D, Giugni JC, Almeida L, et al. Association between antidepressants and falls in Parkinson's disease. *Journal of Neurology* 2016;**263**(1):76–82.
109. Askari M, Eslami S, Scheffer AC, et al. Different risk-increasing drugs in recurrent versus single fallers: are recurrent fallers a distinct population? *Drugs and Aging* 2013;**30**(10):845–51.
110. Parashos SA, Wielinski CL, Giladi N, Gurevich T. Falls in Parkinson disease: analysis of a large cross-sectional cohort. *Journal of Parkinson's Disease* 2013;**3**(4):515–22.
111. Griffin 3rd CE, Kaye AM, Bueno FR, Kaye AD. Benzodiazepine pharmacology and central nervous system-mediated effects. *The Ochsner Journal* 2013;**13**(2):214–23.
112. Rodriguez-Oroz MC, Zamarbide I, Guridi J, Palmero MR, Obeso JA. Efficacy of deep brain stimulation of the subthalamic nucleus in Parkinson's disease 4 years after surgery: double blind and open label evaluation. *Journal of Neurology Neurosurgery and Psychiatry* 2004;**75**(10):1382–5.
113. Micheli F, Cersosimo MG, Piedimonte F. Camptocormia in a patient with Parkinson disease: beneficial effects of pallidal deep brain stimulation. Case report *Journal Of Neurosurgery* 2005;**103**(6):1081–3.
114. Yamada K, Goto S, Matsuzaki K, et al. Alleviation of camptocormia by bilateral subthalamic nucleus stimulation in a patient with Parkinson's disease. *Parkinsonism and Related Disorders* 2006;**12**(6):372–5.
115. Umemura A, Oka Y, Ohkita K, Yamawaki T, Yamada K. Effect of subthalamic deep brain stimulation on postural abnormality in Parkinson disease. *Journal of Neurosurgery* 2010;**112**(6):1283–8.

116. Shih LC, Vanderhorst VG, Lozano AM, Hamani C, Moro E. Improvement of pisa syndrome with contralateral pedunculopontine stimulation. *Movement Disorders: Official Journal of the Movement Disorder Society* 2013;**28**(4):555–6.
117. Ricciardi L, Morgante L, Epifanio A, et al. Stimulation of the subthalamic area modulating movement and behavior. *Parkinsonism and Related Disorders* 2014;**20**(11):1298–300.
118. Nantel J, McDonald JC, Bronte-Stewart H. Effect of medication and STN-DBS on postural control in subjects with Parkinson's disease. *Parkinsonism and Related Disorders* 2012;**18**(3):285–9.
119. Rocchi L, Chiari L, Cappello A, Gross A, Horak FB. Comparison between subthalamic nucleus and globus pallidus internus stimulation for postural performance in Parkinson's disease. *Gait and Posture* 2004;**19**(2):172–83.
120. Tassorelli C, De Icco R, Alfonsi E, et al. Botulinum toxin type A potentiates the effect of neuromotor rehabilitation of Pisa syndrome in Parkinson disease: a placebo controlled study. *Parkinsonism and Related Disorders* 2014;**20**(11):1140–4.
121. Capecci M, Serpicelli C, Fiorentini L, et al. Postural rehabilitation and kinesio taping for axial postural disorders in Parkinson's disease. *Archives of Physical Medicine and Rehabilitation* 2014;**95**(6):1067–75.
122. Bartolo M, Serrao M, Tassorelli C, et al. Four-week trunk-specific rehabilitation treatment improves lateral trunk flexion in Parkinson's disease. *Movement Disorders: Official Journal of the Movement Disorder Society* 2010;**25**(3):325–31.
123. Santamato A, Ranieri M, Panza F, et al. Botulinum toxin type A and a rehabilitation program in the treatment of Pisa syndrome in Parkinson's disease. *Journal of Neurology* 2010;**257**(1):139–41.
124. Kataoka H, Ikeda M, Horikawa H, Ueno S. Reversible lateral trunk flexion treated with a rehabilitation program in a patient with Parkinson's disease. *Parkinsonism and Related Disorders* 2013;**19**(4):494–7.
125. Shen X, Mak MK. Repetitive step training with preparatory signals improves stability limits in patients with Parkinson's disease. *Journal of Rehabilitation Medicine* 2012;**44**(11):944–9.
126. Jobges M, Heuschkel G, Pretzel C, Illhardt C, Renner C, Hummelsheim H. Repetitive training of compensatory steps: a therapeutic approach for postural instability in Parkinson's disease. *Journal of Neurology Neurosurgery and Psychiatry* 2004;**75**(12):1682–7.
127. Qutubuddin AA, Cifu DX, Armistead-Jehle P, Carne W, McGuirk TE, Baron MS. A comparison of computerized dynamic posturography therapy to standard balance physical therapy in individuals with Parkinson's disease: a pilot study. *NeuroRehabilitation* 2007;**22**(4):261–5.
128. King LA, Salarian A, Mancini M, et al. Exploring outcome measures for exercise intervention in people with Parkinson's disease. *Parkinson's Disease* 2013;**2013**:572134.
129. Jeka JJ, Lackner JR. Fingertip contact influences human postural control. *Experimental Brain Research* 1994;**100**(3):495–502.
130. Franzen E, Paquette C, Gurfinkel V, Horak F. Light and heavy touch reduces postural sway and modifies axial tone in Parkinson's disease. *Neurorehabilitation and Neural Repair* 2012;**26**(8):1007–14.

CHAPTER

4

How are postural responses to external perturbations affected by PD?

Clinical case

Mary noticed that she started falling at home about the time when a neurologist noticed that she could not maintain her balance in response to a backward pull on her shoulders. She noticed that she often fell when the environmental conditions suddenly changed. Unfortunately, when she lost her balance, Mary often injured herself as she could not catch herself with a step or her hands. Although she responded very well to levodopa with less rigidity and faster walking, she felt unsteady and fell often. Many of her Parkinson disease (PD) symptoms responded well to Deep brain stimulation (DBS) in the STN but felt her balance was even worse than before surgery. Mary started falling less after practicing improving her balance stepping responses with her physical therapist.

A. How are retropulsion and propulsion related to impaired postural responses?

Retropulsion and propulsion, or multiple small steps in response to backward or forward disequilibrium, respectively, are typical balance impairments that occur as parkinsonism progresses. In fact, progression of PD is marked by inability to recover from a backward pull on the shoulders ("Pull Test") and this test has been embedded in the MDS-UPDRS Motor Part III and causes progression on the Hoehn and Yahr Scale. Fig. 4.1A shows a patient with PD retropulsing with many

Balance Dysfunction in Parkinson's Disease
https://doi.org/10.1016/B978-0-12-813874-8.00004-0

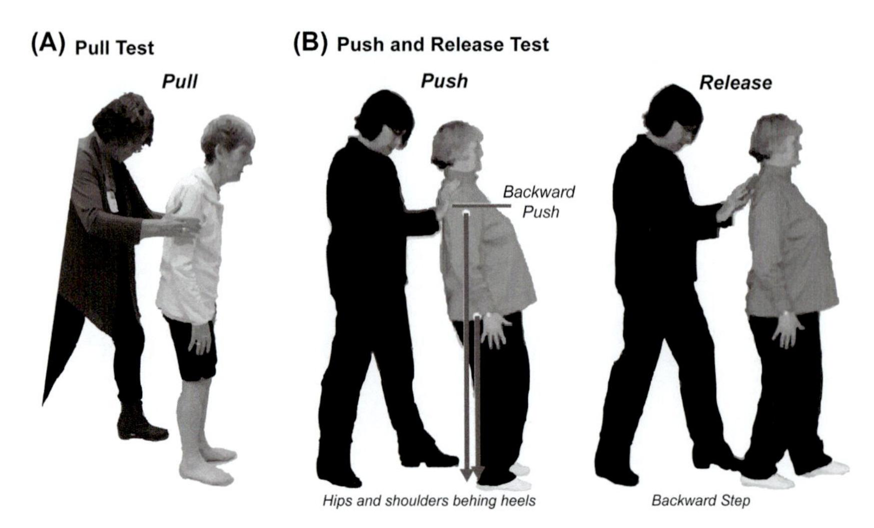

FIGURE 4.1 (A) Photograph of a clinician administering the Pull Test from the Motor UPDRS. (B) Photograph of a clinician administering the backward Instrumented Push and Release Test in a patient with PD who has retropulsion. When the patient is leaning against the clinician's hands, their body CoM is behind their feet and the clinician is supporting their body weight. After the "release," the patient takes a backward step to recover equilibrium, or falls into the clinician's hands.

small steps in response to a backward pull on the shoulders. Because the ability to consistently pull the body Center of Mass (CoM) backward just the amount required for a single step is difficult; with poor interrater and test-retest reliability, we developed an alternative "Push and Release Test." The Push and Release Test involves having the patient lean back (not pushing back) into the examiner's hands just until body weight is supported (i.e., the limits of stability when the body CoM is just over the edge of the base of support). Fig. 4.1B shows a participant recovering with one step in response to a sudden release of upper trunk support during the Push and Release Test.

The Push and Release Test better predicts future falls in people with PD better than the Pull Test.[1] The Push and Release was also shown to be more reliable and sensitive than the Pull Test when administered in the same patients with PD by the same examiners.[2] That is, the forces associated with release were more consistent trial-by-trial within a subject than the forces associated with a Pull Test and more patients showed abnormal responses in the Push and Release Test than in the Pull Test. The Push and Release Test can also be performed in the forward and sideways directions, unlike the Pull Test.[2] The better performance of the Push and Release compared to the Pull Test may be because gravity provides the destabilizing force, not the examiner, the amount of lean is normalized to each subject's body biomechanics (weight, size of base of support, etc.), and it is more difficult for patients to anticipate when the disequilibrium occurs. Recently, we have developed an instrumented version of the Push and Release Test (IPush) using body-worn inertial sensors and validated the measures with laboratory gold-standard technologies.[3] This IPush test provides clinicians with object measures of latency of stepping onset, size of first step, number of steps, and time to recover equilibrium.[3] We are finding that this wearable technology version of manually induced postural responses is sensitive to PD similar to laboratory studies showing normal latencies but multiple short steps resulting in longer time to recover equilibrium. In addition, PD with freezing of gait shows even smaller steps and longer time to recover equilibrium[4] (for more details see Chapter 8).

The normal response to a sudden, large forward or backward postural perturbation is to exert torque into the ground with the ankles (ankle strategy) and then to add a stepping strategy.[5] Sideways perturbations likewise start with postural responses with the feet-in-place and, if large enough, result in sideways stepping responses, either by widening stance or by stepping over a foot.[6] When the body CoM is perturbed beyond the base of foot support, ankle torque, alone, can no longer recover equilibrium and the base of support needs to be moved under the falling CoM.

Studies have shown that both the ankle strategy and the stepping strategy are slow to develop force and weak in PD, but with normal latencies.[6,7] A timely, but slow and weak stepping, response that results in repeating inadequate steps to arrest the falling body CoM would cause retropulsion (backward stepping while falling) or propulsion (forward stepping while falling). Retropulsion normally occurs while standing, either because of spontaneous backward leaning as common with Progressive Supranuclear Palsy, or because of an external perturbation, such as a backward tug on the shoulders. Propulsion often occurs when walking as the body CoM is naturally forward of the feet while walking.

B. Why do patients with PD show more abnormalities in their postural responses in the backward direction than other directions (or do they?)

The Pull Test has been performed for many years only in the backward direction because clinicians find it useful to identify abnormal postural responses. However, some patients with PD show propulsion (i.e., repetitive forward stepping) when walking and most patients with PD show difficulty recovering from a sideways fall with a stepping response. The forward stepping response is more difficult to test in the clinic because the forward stepping response using the Pull Test requires more force and allows more patient anticipation than a backward pull. The sideways stepping response is more difficult to test than the backward or forward stepping because a sideways Push and Release (holding body support at the pelvis before the release) can be dangerous in clinical conditions without a belt and strong examiner, since patients with PD often fail to recover and must be caught (see below).

We demonstrated that feet-in-place postural responses are, indeed, more abnormal in the backward direction, than other directions of postural disequilibrium in patients with PD.[8] Specifically, the percentage of trials with falls in response to surface translations were greatest for the backward (and backward left) falling directions (Fig. 4.2A). It is not clear why falls to the left resulted in more falls than falls to the right in this cohort, but our assistant was standing to the left of subjects, which may have influenced perceived safety in fall direction.

The stability margin in response to small perturbations was smallest for the backward direction in subjects with PD, but not in elderly control subjects. We used the area of the difference between the CoP and CoM displacement (Fig. 4.3) to plot the relative stability limits in response to perturbations across eight different directions (Fig. 4.2B). The age-matched control subjects showed similar limits of stability to perturbations across all eight directions (*pink circle* in Fig. 4.2B). In contrast, the

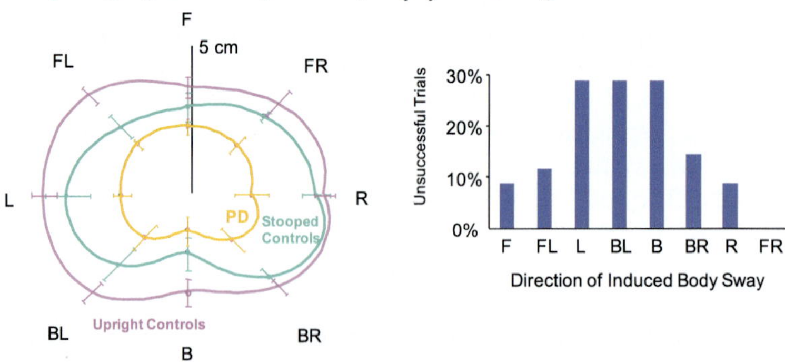

FIGURE 4.2 Postural responses are more abnormal backward than other directions. (A) Stability margin (difference between peak CoP and peak CoM) in response to surface translations (9 cm displacement and 1 s ramp duration) across eight directions (*F*, forward; *FR*, forward-right; *R*, right; *BR*, backward-right; *B*, backward; *BL*, backward-left; *L*, left; *FL*, forward left) in seven subjects with moderate to severe PD (*yellow*), and seven age-matched control subjects standing naturally (*pink*) and standing with a flexed posture to imitate PD (*blue*). (B) The percent of 10 perturbation trials with a fall across eight directions in subjects with PD shows the most falls in the backward direction *From Horak FB, Dimitrova D, Nutt JG. Direction-specific postural instability in subjects with Parkinson's disease.* Experimental Neurology *2005;**193**(2): 504–521.*

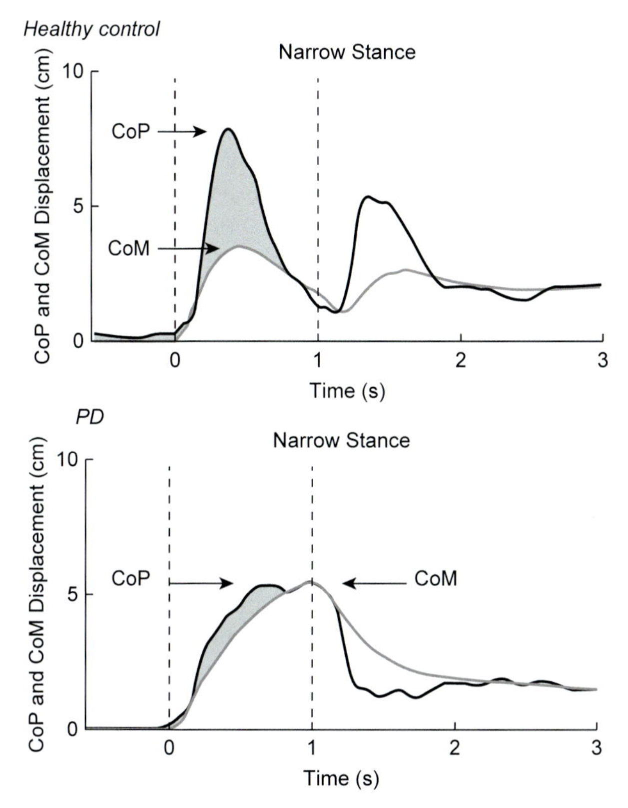

FIGURE 4.3 Representative example of bradykinetic, but with normal latency, postural response in a subject with PD compared to a healthy control subject. The CoP response to the backward surface translation is much larger in the control than the response in the subject with PD, resulting in smaller displacement of the body CoM in the control than the PD subject. Difference between the CoM and CoP displacement represented by gray shading shows that the subject with PD would likely have fallen if not for the deceleration provided by the stop of the platform translation (*second dashed line*). *Adapted from Horak FB, Dimitrova D, Nutt JG. Direction-specific postural instability in subjects with Parkinson's disease.* Experimental Neurology *2005;**193**(2):504–521.*

subjects with PD showed much smaller limits of stability, particularly in response to perturbations that resulted in backward postural sway. When control subjects were asked to assume the flexed posture of patients with PD, their limits of stability shrunk toward that of subjects with PD, especially in response to backward sway perturbations (Fig. 4.2B). The effect of flexed postural alignment on postural stability in response to backward sway perturbations is therefore consistent with a biomechanical explanation for why patients with PD are so vulnerable to falls in response to backward perturbations. Perhaps a shortened tibialis muscle in flexed posture has difficulty developing adequate torque to resist backward sway. This improvement in postural stability in response to external perturbations when subjects improve their postural alignment also highlights the importance of rehabilitation focused on improving postural alignment to prevent falls.

In addition to feet-in-place postural responses, stepping responses are also more abnormal in the backward, than forward, direction in patients with PD.[9] Differences in stepping length are greater between PD and age-matched control subjects in the backward, than forward, direction, resulting in larger body CoM displacements backwards, than forward. However, biomechanics can also contribute to direction-specific, postural instability because the backwards base of foot support are much smaller than the forward or sideways limits of stability so the body CoM more quickly exceeds the limits and thus, requires a bigger and faster step.

Backward stepping responses may be more affected than forward responses because subjects cannot see their legs while stepping backward. Subjects with PD may rely more on vision to control their postural stepping responses than control subjects because of impaired kinesthesia. Many studies have demonstrated impaired central perception of kinesthesia in patients with PD, including joint position sense, sense of weight, and pressure/touch.[10] *Kinesthesia* refers to the perception of the position and movement of one's own body, limbs, and muscles. We showed that when patients with PD could see both their feet and a target on the floor, they could significantly increase the size of their forward stepping response to a backward surface displacement.[11] However, if subjects with PD could see the target, but not their leg, they significantly undershoot the target with small stepping responses. Control subjects reached the target on the floor with their externally triggered stepping response, whether or not they could see their leg, consistent with use of kinesthesia to appropriately scale the size of their step.

C. What are the major postural responses deficits in PD?

The most obvious abnormality of postural responses to external perturbations in PD is smaller than normal strength of postural responses. For example, Fig. 4.3 shows smaller peak CoP and later time to peak CoP displacement in response to a backward perturbation, resulting in forward falling in a subject with PD compared to a control subject. As a consequence of slow, weak development of forces resisting the perturbations, the body CoM falls farther and faster in subjects with PD than in control subjects.

PD is associated with worse stability in response to perturbations in all directions, but the difference is largest for the backward direction.[12] The closer the peak body CoM approaches the CoP displacement resisting it, the more likely a fall. Fig. 4.2A shows a large difference in CoP and CoM trajectory in the control, but not the subject with PD in response to backward disequilibrium.

Surprisingly, despite the slowness to develop force and lower peak forces of postural responses, postural response latencies are normal in patients with PD, even as the disease progresses with more bradykinesia and more falls. The normal latencies of postural responses, known to be triggered by proprioceptive stimulation, suggest that the dopaminergic basal ganglia networks and the cholinergic networks affected by late stage PD are not critical for triggering the initial latencies of postural responses. Also, these results tell us that bradykinesia and akinesia do not affect the speed of onset of postural responses, although they greatly affect the rate of development and peak muscle torque exerted by postural responses.[13]

Muscle activation patterns in response to postural responses are also abnormal in patients with PD. Both control subjects and subjects with PD activate a distal to proximal pattern of muscle activation when a feet-in-place ankle strategy is used. The latency of ankle muscle synergy activation does not differ between PD and control subjects, with activation of ankle muscles at about 100 ms, followed by knee, and then hip muscles when an ankle strategy is used (see Fig. 4.4). However, unlike control

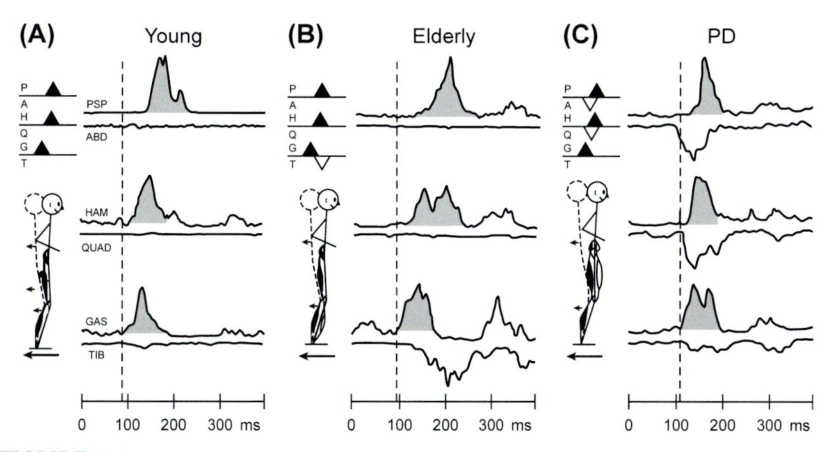

FIGURE 4.4 Typical patterns of muscle activation in a representative young (A), elderly (B), and subject with PD in the OFF levodopa state (C). Dorsal muscle bursts associated in the ankle strategy in response to backward surface translations (forward falls) are shaded gray and ventral muscle bursts associated with a hip strategy are thickened. The EMG activity is filtered and rectified and aligned to the time of onset of the surface translation (0 ms). The dashed vertical line shows the latency of the earliest muscle burst in the gastrocnemius muscle. Stick figures and triangles summarize muscle activation patterns and small arrows indicate the direction of active postural correction for forward sway displacements. *Adapted from Horak FB, Nutt JG, Nashner LM. Postural inflexibility in parkinsonian subjects.* Journal of the Neurological Sciences *1992;**111**(1):46–58.*

subjects, people with PD often add bursts of muscle activation in antagonist muscles resulting in coactivation that would stiffen joints. When people with PD take levodopa medication, however, antagonist muscle activation disappears, so postural response muscle activation temporal patterns resemble normal and joints are less rigid, although the magnitude of postural responses is not increased to normal by levodopa.

Responses to large forward perturbations result in use of hip, trunk, and knee motion as well as ankle joint motion to return the body to equilibrium quickly. Whereas control subjects flex the hips/trunk in response to sideways perturbations and flex the knees in response to backward perturbations, people with PD keep their joints rigid. This rigidity results in larger body CoM displacements to the same perturbations. Fig. 4.5A compares hip flexion in response to lateral surface translations in control subject with a subject who has PD and Fig. 4.5B compares knee flexion in response to a backward surface translation in the same subjects. This rigidity of joint motion in response to external displacements results in increased displacement of the body CoM (i.e., falling farther) in Fig. 4.2. Stiffness of postural responses in patients with PD have been shown to be due to more cocontraction activity of agonist and antagonist muscles at both medium (80 ms) and balance correcting (120 ms) response latencies and increased background activity in lower leg, hip, and trunk muscles.[12]

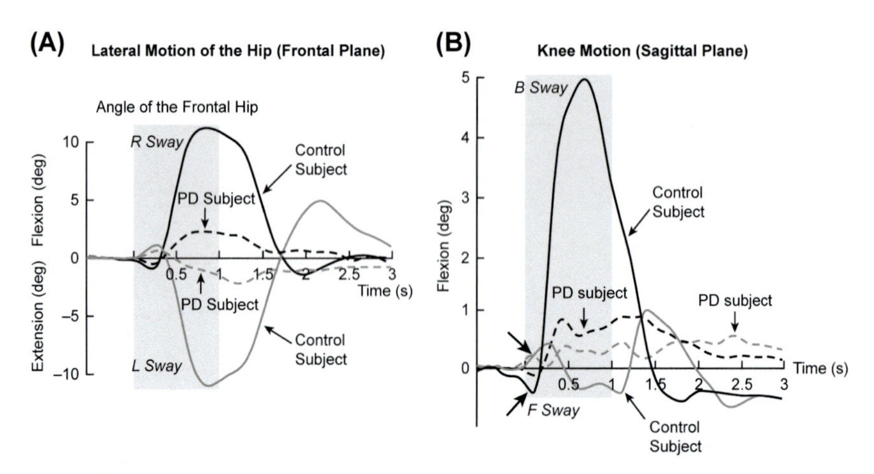

FIGURE 4.5 Subject with PD shows smaller than normal (control subject) right and left hip/trunk range and knee flexion in response to lateral surface perturbations. (A) Frontal plan hip lateral flexion in a representative control subject and subject with PD in response to right and left perturbations. (B) Sagittal plane knee flexion in a representative control subject and subject with PD in response to the same perturbations. *Adapted from Horak FB, Dimitrova D, Nutt JG. Direction-specific postural instability in subjects with Parkinson's disease.* Experimental Neurology *2005;**193**(2):504–521.*

Rigidity of joint motion as well as impaired kinesthesia may also underlie the lack of protective arm reaching responses to postural perturbations. Healthy subjects quickly and automatically flex or extend and abduct their shoulders so their arms lead their trunks when their postural stability is perturbed, likely to catch themselves on the floor in case their postural responses are unsuccessful. Although the patients had significantly earlier onset of deltoid muscle responses, this gave no functional protection because the arm movements were abnormally directed, slower velocity and smaller peak displacements, particularly in the OFF state.[12] In fact, patients with PD often adducted their arms, instead of abducting them with high variability of arm postural responses.[12] Impaired protective arm postural responses could help explain why patients with PD more often fracture or injure themselves in falls than elderly people without PD.[14]

D. How does PD affect adaptation of postural responses to changes in environmental context?

The inability to quickly change postural response strategy when current conditions change may be a unique sensorimotor control problem in people with PD that reflects an important role of the basal ganglia. Although postural responses are triggered quickly and automatically, they must change with changes in physical or mental conditions. Whereas trial-by-trial adaptation based on prior experience appears to require an intact cerebellum,[15,16] first trial changes immediately after a change in conditions appears to require intact basal ganglia. The first published example of this lack of flexibility in postural response synergies is shown in Fig. 4.6. Whereas healthy subjects quickly change from using a distal-to-proximal ankle synergy when responding to surface translations on a flat surface, they immediately begin to change to use of a proximal hip synergy when standing on the beam and then to use a trunk only synergy when it would be ineffective to activate leg muscles when the legs are dangling while sitting on a stool. Patients with PD, whether in the ON or OFF state, initially show an inflexible strategy regardless of their surface support conditions, with a gradual change with many repetitions. This inability to quickly change "postural set" such that postural response synergies to the same perturbations are altered based on initial physical conditions is inefficient and could lead to falls during transitions from one condition to another. This inflexibility also explains why patients with PD tend not to improve as much as expected with manual support and find it difficult to benefit from canes and walkers without much practice.

Normally, the magnitude of postural responses can be greatly influenced by a subject's intent, or mental "central set." For example, the size

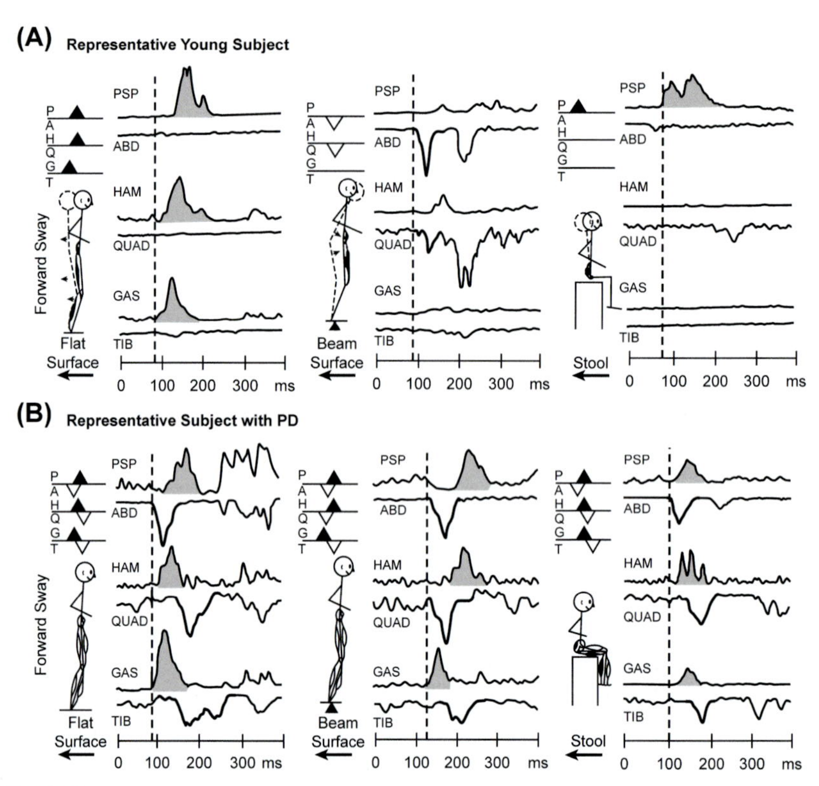

FIGURE 4.6 Subjects with PD do not show adaptation of postural responses for changes in support surface conditions. EMG responses to backward surface displacements in a representative control subject and subject with PD under three different support surface conditions: flat surface, narrow beam, and sitting with leg dangling. The control subject changes from use of an ankle synergy on the flat surface to a hip synergy on the narrow beam to trunk muscles only when sitting. In contrast, the subject with PD reduces the amplitude but does not alter postural muscle response synergy with changes in support surface conditions, using muscle coactivation across the ankle, knee, and hip joints. *Adapted from Horak FB, Nutt JG, Nashner LM. Postural inflexibility in parkinsonian subjects.* Journal of the Neurological Sciences *1992;**111**(1):46–58.*

of gastrocnemius muscle activation and its accompanying surface reactive torque can be increased when instructed to "resist" the upcoming surface translation and decreased when instructed to "give" into the perturbation and let yourself be moved by it (Fig. 4.7A). This flexibility in size of postural response when intention changes allows the nervous system to modify the size of its response to an externally triggered postural response based on intention. For example, if a subject anticipated that allowing excessive body motion in response to an upcoming slip or trip could lead to a fall or injury on ice, responses could be "tuned up" to be larger than usual. In contrast, the size of the postural response, both

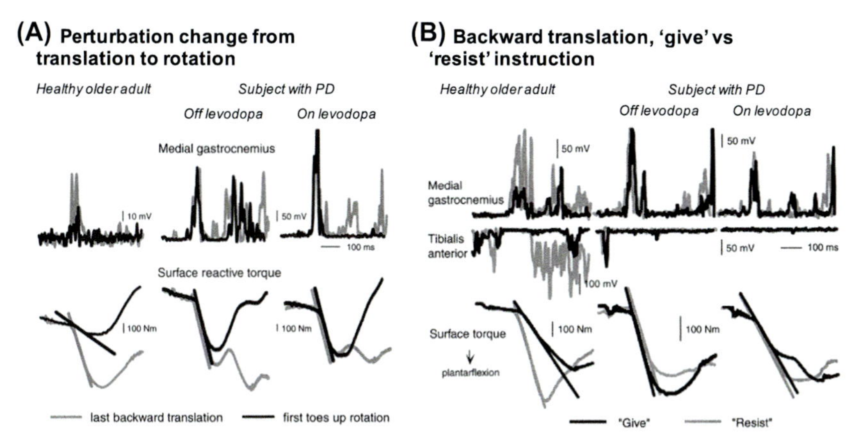

FIGURE 4.7 Patients with PD do not show adaptation of postural responses for changes in context. (A) Control subject, but not subject with PD in the OFF or ON levodopa state immediately change the size of their postural response (gastrocnemius EMG and resulting surface reactive torque) when the perturbation changes from a translation to a rotation. (B) Control subject, but not subject with PD in the OFF or ON levodopa state immediately change the size of their postural response (gastrocnemius EMG and resulting surface reactive torque) when instructed to "give" versus "resist" in response to the surface translations. *Adapted from Chong RK, Horak FB, Woollacott MH. Parkinson's disease impairs the ability to change set quickly.* Journal of the Neurological Sciences *2000;***175***(1):57–70.*

muscle activation and resulting torque, are not altered by instructions to people with PD to either "resist" or "give" into the upcoming postural perturbation.[17] This inflexibility to "mental set" makes it difficult for people with PD to increase the size of their postural responses simply by "thinking big" and makes it difficult for them to decrease the size of their postural response to allow more postural sway without stiffening, which is efficient for the situation.

E. Does levodopa improve postural responses?

Surprisingly, levodopa replacement therapy does not improve, and may further impair, postural response strength, although it reduces bradykinesia of voluntary movement and gait. The ability to quickly increase force on the ground in response to external surface perturbations weakens even more when patients with PD are in their ON levodopa state, compared to their OFF state. Fig. 4.8 compares postural responses to 3 cm and 1.2 cm surface translations in a control subject (left column) with a subject with PD OFF (middle column) and with the same subject with PD ON (right column). The slope showing the rate of change of force is less than one half of the control subject when in the OFF state and then is

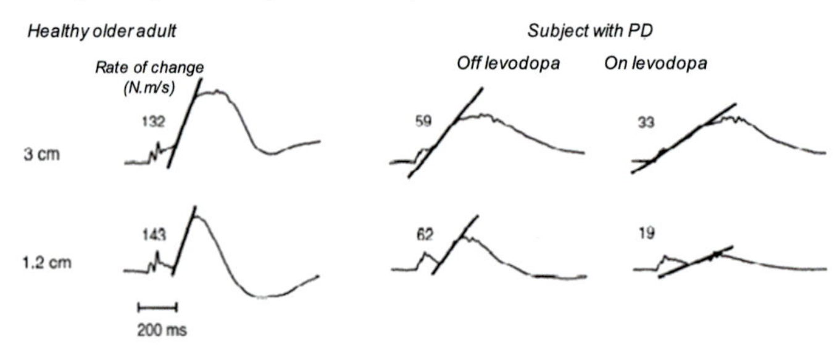

FIGURE 4.8 Postural responses are smaller in a subject with PD than in the control subject, particularly when the subject with PD is in the ON levodopa state. Center of Pressure (CoP) displacement in response to backward surface translations of 3 cm and 1.2 cm in a representative control subject (*left*), and a subject with PD in the OFF and ON levodopa state (middle and right column, respectively). The slope of the initial rate of change of postural response in newton*meters/seconds is shown. *Adapted from Horak FB, Frank J, Nutt J. Effects of dopamine on postural control in parkinsonian subjects: scaling, set, and tone.* Journal of Neurophysiology *1996;75(6):2380–2396.*

weakened by 50% more when the subject takes his levodopa. Thus, it is not surprising that patients with PD fall often, even when they are in their optimal ON state.[18] In fact, they may fall more in their ON, than OFF state, since they move more quickly with reduction of bradykinesia and occasional dyskinesia, show reduced rigidity which allows them to fall more quickly, and do not improve or may weaken their postural response magnitude with levodopa.

The scaling of the amplitude of postural responses to increasing size or velocity of postural perturbations is also not improved by levodopa. In addition to the size of postural response being reduced by levodopa, postural response scaling is further impaired by levodopa. Fig. 4.9A shows the group mean amplitude of postural responses to five increasing amplitudes of postural responses in elderly control subjects and the same subjects with PD in the OFF and ON levodopa state. Whereas the elderly control group continues to scale up the size of their postural response with the increasing amplitudes of postural perturbations, the PD group scales up their responses only for the smaller amplitude perturbations, particularly when in the ON levodopa state. Thus, differences in the size of postural responses between the PD and control group are exaggerated for the largest postural perturbations because of failure of the PD group to sufficiently increase the size of their responses to larger perturbations, especially when they are in the ON state.

Although levodopa does reduce cocontraction in postural muscle synergies, levodopa therapy does not improve the ability to adapt postural synergies when the support surface conditions change. Fig. 4.9B illustrates improvement in the postural response synergy when a

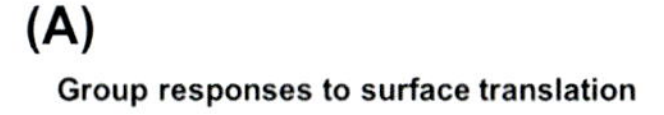

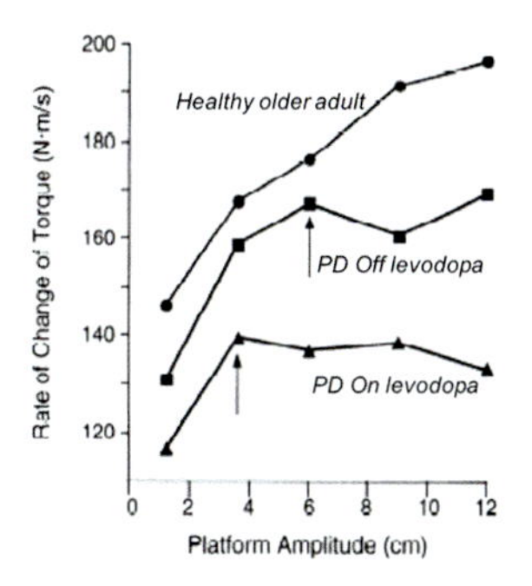

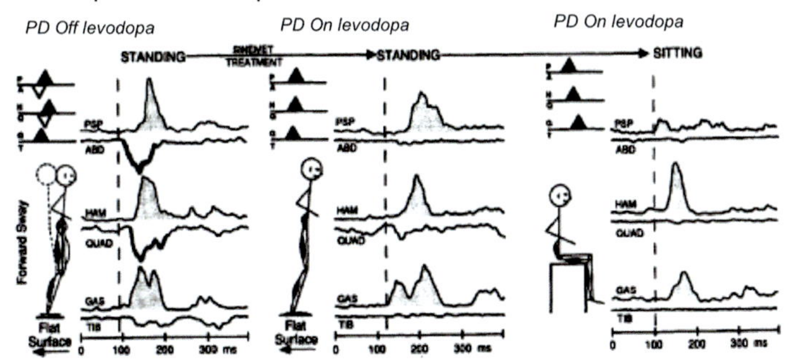

FIGURE 4.9 Levodopa impairs adaptation of postural responses. A. Initial rate of change of postural torque response to five different amplitudes of surface translation in a group of 10 control subjects and 10 subjects with PD in the ON and OFF levodopa state. Inability to scale up the size of postural responses based on experience with larger perturbations is worse in the ON than OFF levodopa state. B. Inability of levodopa to improve adaptation of postural muscle response synergy with changes in support surface conditions from a flat surface to sitting on a stool with legs dangling, despite improvement in the pattern of the postural synergy itself with reduced muscle cocontraction. *Adapted from Horak FB, Nutt JG, Nashner LM. Postural inflexibility in parkinsonian subjects.* Journal of the Neurological Sciences *1992;**111**(1): 46–58; Horak FB, Frank J, Nutt J. Effects of dopamine on postural control in parkinsonian subjects: scaling, set, and tone.* Journal of Neurophysiology *1996;**75**(6):2380–2396.*

representative subject with PD takes their levodopa medication but inability to inhibit leg muscle activation when responding to surface perturbations when sitting on a stool with legs dangling. Thus, the spatial-temporal design of postural synergies is influenced by levodopa, likely related to reduction of rigidity. However, inability to adapt or modify postural responses based on initial conditions is dopa-sensitive, so will impair the ability to respond effectively to a slip or trip, even when in an optimal levodopa state.

It is not known why levodopa worsens automatic postural responses. Voluntary movement force is improved by levodopa[19] but postural responses must use different central circuitry than voluntary muscle force (Chapter 1). Parkinsonian rigidity that stiffens joints and leads to instability (Fig. 4.5) is improved by levodopa. Muscle cocontraction that would reduce effective torques into the support surface is reduced by levodopa so this should lead to more forceful responses. In fact, this reduced stiffness does increase the speed of displacement of the body CoM[7] but does not increase the speed of response to compensate for the faster falling. Perhaps levodopa replacement has a negative effect on other neurotransmitters important for postural response circuitry (Chapter 1). The presence of extraneous levodopa has been shown to reduce cortical cholinergic activity and cholinergic pathways critical for postural control from the Pedunculopontine and the Nucleus Basalis of Meynert, nuclei known to be affected by PD (Chapter 2).

F. Does deep brain stimulation improve postural responses?

Although deep brain stimulation (DBS) in the STN or GPi improves many motor symptoms of PD, we have found that DBS does not improve, and may actually further impair postural responses.[20] DBS stimulation in STN, in particular, results in worsening of postural responses to surface translations 6 months after surgery. In contrast to DBS in the STN, DBS in the GPi for PD does not significantly impair or improve postural responses. Fig. 4.10A summarizes postural stability in response to backward surface translations in the OFF and ON levodopa states prior to DBS surgery and after surgery, with or without DBS turned on.

Subjects with PD prior to surgery in the OFF or ON levodopa states showed worse postural stability than age-matched controls. Although turning the DBS current ON improved postural responses stability for both STN and GPi sites, the negative effect of DBS surgery itself for the STN group was greater than the benefit of the stimulating current, making overall postural responses stability functionally worse after surgery for the STN group. Turning the DBS on at either the STN or GPi site improved postural responses compared to the postoperative OFF condition by lengthening the tibialis response, whereas medication did not show an appreciable effect. This result suggests that DBS can affect non-dopaminergic networks that are involved in postural stabilization. However, the STN group had worse postural responses in their best functional state (DBS + DOPA) 6 months after the DBS procedure compared to their best functional state (ON levodopa) before the DBS procedure. The deficit at 6 months testing in the STN group cannot be accounted for by a change in levodopa dose, as this did not change

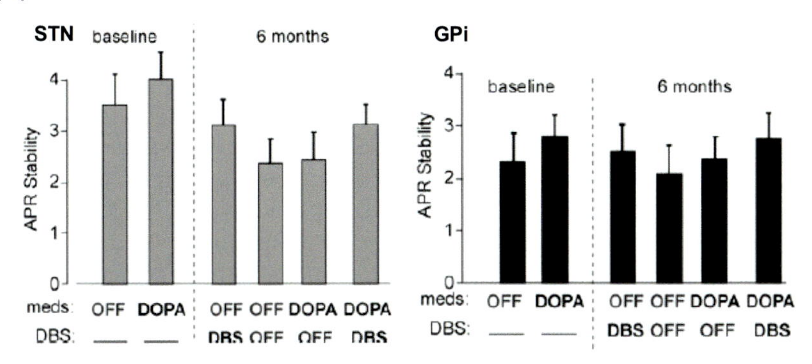

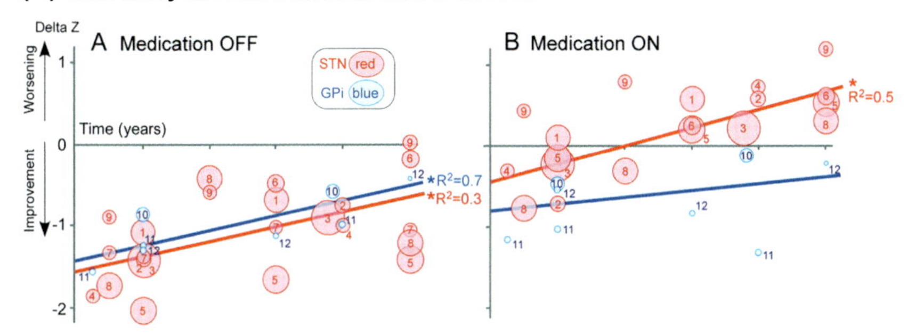

FIGURE 4.10 (A) Worsening of postural responses after DBS surgery in STN but not GPi. Bar graphs show postural stability (area between the CoP and CoM curves as in Fig. 4.3) in response to backward surface translations in the OFF and ON levodopa state prior to surgery and in four conditions after surgery: OFF levodopa/ON DBS, OFF both levodopa and DBS, ON levodopa/OFF DBS, and ON both levodopa and DBS. Larger stability during automatic postural responses indicates improved postural responses (age-matched normal mean Stability APR is around 6). (B) Metaanalysis of 11 papers comparing the effects of DBS in STN and in GPi on the postural instability and gait disability (PIGD) subscore from the Motor UPDRS. Both DBS in STN and GPi result in short-term improvement of PIGD which worsens over time. In the OFF levodopa state, both the STN and GPi groups are back to baseline levels of PIGD in 5 years. In contrast, in the ON levodopa state, the STN group shows faster progression of PIGD than GPi, and becomes worse than baseline by 2 years. *(A) Adapted from St George RJ, Carlson-Kuhta P, Burchiel KJ, Hogarth P, Frank N, Horak FB. The effects of subthalamic and pallidal deep brain stimulation on postural responses in patients with Parkinson disease.* Journal of Neurosurgery *2012;**116**(6):1347–1356; (B) Adapted from St George RJ, Nutt JG, Burchiel KJ, Horak FB. A meta-regression of the long-term effects of deep brain stimulation on balance and gait in PD.* Neurology *2010;**75**(14):1292–1299.*

significantly pre-to postsurgery. Postural responses stability impairment in the STN group was associated with smaller tibialis amplitudes, but no change in response latency or coactivation with gastrocnemius. The STN group was worse after the surgery due to a worsening in postural bradykinesia. This result may account for the increased number of falls reported in PD subjects with DBS in STN than in GPi.[22]

Results from this laboratory study are somewhat consistent with the literature on the effects of DBS in STN and GPi on clinical rating scales of balance and gait. For example, a metaanalysis of studies evaluating the effects of DBS in STN or GPi on postural instability and gait disability (PIGD) before and after surgery show faster progression of PIGD in the STN than the GPi group when evaluated in the ON levodopa state[21] (Fig. 4.10B). However, the PIGD rating scale combines scores for postural responses (tug backward on the shoulders) as well as gait, rise from a chair and postural alignment, each domain of postural control that likely involves separate neural networks (Chapter 1).

G. Does rehabilitation improve postural responses?

Although automatic postural responses are triggered quickly and automatically, several studies have shown they can be improved with practice in people with PD, as well as other balance disorders.[9,23] A recent review of nine studies of training postural responses in older adults showed improved responses after 1 day of training in eight studies with several studies reporting a reduction of falls after training.[23] Feet-in-place postural responses may improve with practice as well in people with PD and controls whereas postural stepping responses improve better in controls than in people with PD.

For example, our recent study showed that a majority of participants with PD improved their postural responses with only 1 day of practice as well as controls standing on a forward and backward oscillating surface.[24] Improvements in postural responses were revealed by reduced body CoM displacements (gain) and improved phase relationships between the CoM and platform motion (Fig. 4.11A). Rates of improvement were comparable between the PD and control groups, demonstrating preserved adaptive capacity for participants with PD. Similar to control participants, the PD group moved toward anticipatory CoM control as a strategy for improving stability, although they never actually anticipated platform displacements like the control group. The PD group exhibited short-term retention of performance improvements the next day and demonstrated generalizability of the learned responses to a different oscillating pattern as well as the control group. Rate of improvement with practice, but not retention, was related to severity of PD motor impairments from the UPDRS Motor score.

One day of practice can also improve postural stepping responses in people with PD. However, unlike control subjects, improvements in people with PD occurred primarily in the first block of trials. In addition,

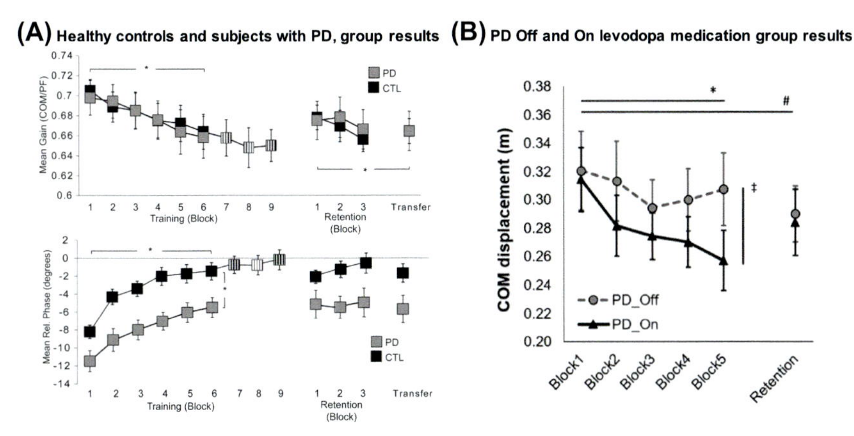

FIGURE 4.11 (A) Improvement of feet-in-place postural responses to continuous surface oscillations with practice in people with PD and age-matched control subjects. The gain (displacement of body CoM/displacement of the support surface) improved similarly with similar retention and transfer in the PD and control groups. The phase (relative time of body CoM displacement reversal compared to time of platform displacement reversal) was more delayed in the PD group with slower learning curve. (B) Improvement of postural stepping responses to surface translations with practice in people with PD. Average values and standard deviation (SD) across five perturbations are shown over five blocks of practice, followed by a retention block the following day. PD subjects improved postural responses with practice more when in the ON levodopa state, than in the OFF levodopa state. *(A) From Van Ooteghem K, Frank JS, Horak FB. Postural motor learning in Parkinson's disease: the effect of practice on continuous compensatory postural regulation.* Gait & Posture *2017;***57***:299–304. (B) From Peterson DS, Horak FB. The effect of levodopa on improvements in protective stepping in people with Parkinson's disease. Neurorehabilitation and Neural Repair 2016;***30***(10):931–940.*

improvements were more pronounced during backward protective stepping than forward stepping, which was not as abnormal as backward stepping. Improvements of postural stepping responses, with the exception of step latency, were retained 24 h later. Unfortunately, improvements in forward–backward stepping did not generalize to lateral protective stepping. Thus, people with PD can improve protective stepping over the course of 1 day of perturbation practice with improvements and retention similar to healthy adults.

Perturbation training appears to be more effective if patients with PD are optimally medicated. Learning to improve postural stepping responses in one session of 25 perturbations is better in the ON levodopa state, compared to the OFF levodopa state (Figure 4.11B).[25] People with PD reduced their CoM displacement and improved margin of stability over training. Improvements in these outcomes were more pronounced after training while ON levodopa than OFF levodopa. Unfortunately, people with PD who freeze exhibited a reduced ability to improve their postural stepping responses.[25]

Technology-assisted balance and gait training over 6 weeks also result in significant improvements in the latency and stride length of compensatory postural stepping responses to forward translations of the support surface compared to an active control group doing strengthening exercises.[26]

Highlights

- Patients with PD have particular difficulty recovering equilibrium in response to backward perturbations.
- Postural responses are weak and slow in PD (bradykinetic) but have normal latency.
- Muscle cocontraction and stiff joints impair postural responses to perturbations in people with PD.
- Inability to quickly change the pattern of postural response to changes in environmental context is unique to PD.
- Levodopa does not improve the strength or adaptation of postural responses.
- DBS in STN further impairs postural responses and DBS in GPi does not improve postural responses.
- Rehabilitation involving practice responding to postural perturbations can improve postural responses in people with PD.

References

1. Hass CJ, Bloem BR, Okun MS. Pushing or pulling to predict falls in Parkinson disease? *Nature Clinical Practice Neurology* 2008;**4**(10):530–1.
2. Jacobs JV, Horak FB, Van Tran K, Nutt JG. An alternative clinical postural stability test for patients with Parkinson's disease. *Journal of Neurology* 2006;**253**(11):1404–13.
3. El-Gohary M, Peterson D, Gera G, Horak FB, Huisinga JM. Validity of the instrumented Push and release test to quantify postural responses in persons with multiple sclerosis. *Archives of Physical Medicine and Rehabilitation* 2017;**98**(7):1325–31.
4. Carlson-Kuhta P, Laird A, Mancini M, et al. *The effect of cognitive dual-task on balance and gait in older men Society for Neuroscience 2016*. 2016. San Diego, California.
5. Maki BE, McIlroy WE. The role of limb movements in maintaining upright stance: the "change-in-support" strategy. *Physical Therapy* 1997;**77**(5):488–507.
6. King LA, Horak FB. Lateral stepping for postural correction in Parkinson's disease. *Archives of Physical Medicine and Rehabilitation* 2008;**89**(3):492–9.
7. Horak FB, Nutt JG, Nashner LM. Postural inflexibility in parkinsonian subjects. *Journal of the Neurological Sciences* 1992;**111**(1):46–58.
8. Horak FB, Dimitrova D, Nutt JG. Direction-specific postural instability in subjects with Parkinson's disease. *Experimental Neurology* 2005;**193**(2):504–21.
9. Peterson DS, Dijkstra BW, Horak FB. Postural motor learning in people with Parkinson's disease. *Journal of Neurology* 2016;**263**(8):1518–29.

10. Konczak J, Corcos DM, Horak F, et al. Proprioception and motor control in Parkinson's disease. *Journal of Motor Behavior* 2009;**41**(6):543–52.
11. Jacobs JV, Horak FB. Abnormal proprioceptive-motor integration contributes to hypometric postural responses of subjects with Parkinson's disease. *Neuroscience* 2006; **141**(2):999–1009.
12. Carpenter MG, Allum JH, Honegger F, Adkin AL, Bloem BR. Postural abnormalities to multidirectional stance perturbations in Parkinson's disease. *Journal of Neurology Neurosurgery and Psychiatry* 2004;**75**(9):1245–54.
13. Horak FB, Frank J, Nutt J. Effects of dopamine on postural control in parkinsonian subjects: scaling, set, and tone. *Journal of Neurophysiology* 1996;**75**(6):2380–96.
14. Pickering RM, Grimbergen YA, Rigney U, et al. A meta-analysis of six prospective studies of falling in Parkinson's disease. *Movement Disorders* 2007;**22**(13):1892–900.
15. Horak FB, Diener HC. Cerebellar control of postural scaling and central set in stance. *Journal of Neurophysiology* 1994;**72**(2):479–93.
16. Timmann D, Horak FB. Prediction and set-dependent scaling of early postural responses in cerebellar patients. *Brain* 1997;**120**(Pt 2):327–37.
17. Chong RK, Horak FB, Woollacott MH. Parkinson's disease impairs the ability to change set quickly. *Journal of the Neurological Sciences* 2000;**175**(1):57–70.
18. Grimbergen YA, Munneke M, Bloem BR. Falls in Parkinson's disease. *Current Opinion in Neurology* 2004;**17**(4):405–15.
19. Teo WP, Rodrigues JP, Mastaglia FL, Thickbroom GW. Modulation of corticomotor excitability after maximal or sustainable-rate repetitive finger movement is impaired in Parkinson's disease and is reversed by levodopa. *Clinical Neurophysiology* 2014;**125**(3): 562–8.
20. St George RJ, Carlson-Kuhta P, Burchiel KJ, Hogarth P, Frank N, Horak FB. The effects of subthalamic and pallidal deep brain stimulation on postural responses in patients with Parkinson disease. *Journal of Neurosurgery* 2012;**116**(6):1347–56.
21. St George RJ, Nutt JG, Burchiel KJ, Horak FB. A meta-regression of the long-term effects of deep brain stimulation on balance and gait in PD. *Neurology* 2010;**75**(14):1292–9.
22. Weaver FM, Follett KA, Stern M, et al. Randomized trial of deep brain stimulation for Parkinson disease: thirty-six-month outcomes. *Neurology* 2012;**79**(1):55–65.
23. McCrum C, Gerards MHG, Karamanidis K, Zijlstra W, Meijer K. A systematic review of gait perturbation paradigms for improving reactive stepping responses and falls risk among healthy older adults. *European review of aging and physical activity* 2017;**14**:3.
24. Van Ooteghem K, Frank JS, Horak FB. Postural motor learning in Parkinson's disease: the effect of practice on continuous compensatory postural regulation. *Gait & Posture* 2017;**57**:299–304.
25. Peterson DS, Horak FB. The effect of levodopa on improvements in protective stepping in people with Parkinson's disease. *Neurorehabilitation and Neural Repair* 2016;**30**(10): 931–40.
26. Shen X, Mak MK. Balance and gait training with augmented feedback improves balance confidence in people with Parkinson's disease: a randomized controlled trial. *Neurorehabilitation and Neural Repair* 2014;**28**(6):524–35.

CHAPTER

5

How are anticipatory postural adjustments in preparation for voluntary movements affected by PD?

Clinical case

Since early in her diagnosis, Mary started experiencing difficulties with initiating locomotor movements, such as rising from a chair and initiating gait (called "start hesitation"). Levodopa therapy helped her tremendously; however, over the years, she experienced OFF-ON fluctuations and found more difficulty to initiate movements. After talking to her doctor who was skeptical about STN or GPi DBS to help gait initiation, Mary consulted her physical therapist. After her first session, she started using with a laser cue to initiate her movements while she feels OFF medication and she experienced a marked improvement.

A. Are APAs hypometric in PD?

Anticipatory postural adjustments (APAs) tend to minimize the potential postural disturbance of forthcoming movements and enable equilibrium to be maintained during movement execution. For example, execution of an arm raising movement in healthy adults is preceded by a sequence of recruitment of the postural (leg and trunk) muscles.[1,2] In people with Parkinson disease (PD), different types of abnormalities concerning APAs associated with voluntary arm movements have been published. Previous reports of anticipatory adjustments in patients with

https://doi.org/10.1016/B978-0-12-813874-8.00005-2

PD have suggested abnormal timing of the electromyographic (EMG) bursts of postural muscles, relative to the prime mover. Specifically, a lack of anticipatory change in the activity of postural muscles during arm movements has been reported for standing subjects.[3,4] In contrast, the presence of APAs in patients has been detected in other studies,[5,6] although certain abnormalities were observed: for example, Dick and colleagues[5] showed that the postural muscle EMG burst amplitude was lower in patients with PD (compared with controls), whereas Rogers and colleagues[6] showed that the recruitment of postural muscles was less frequent and was characterized by multiple EMG bursts extending to the antagonists as well as agonists.

APAs are also associated with lower limb movements. Postural adjustments associated with the task of rising on tip toe were investigated, and results were controversial. For example, in a reaction time paradigm by Diener and colleagues[7] the basic pattern of preparatory and executional activities was preserved in most patients with PD. The time intervals between postural preparation and execution were also normal but these subjects were tested in the ON state. Instead, in the paper from Frank et al.,[8] subjects with PD, OFF their medication, showed reduced magnitudes and delayed timing of the postural and voluntary components of the rise-to-toes task, as if they had difficulty turning off the postural tibialis anterior (TIB) component and initiating the voluntary gastrocnemius (GAS) component, see Fig. 5.1. Dopamine improved the relative timing, as well as the magnitude of both postural and voluntary components of rise-to-toes.[8] Although the magnitude of dorsiflexion torque was smaller in PD compared to healthy controls, the subjects with

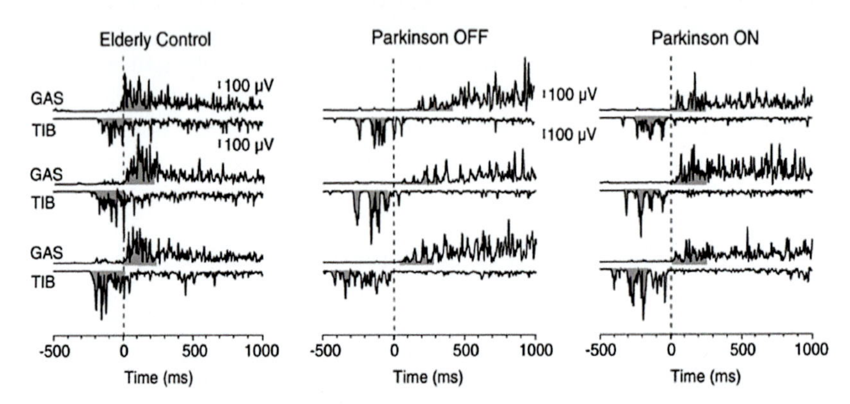

FIGURE 5.1 Longer APA muscle bursts in tibialis (TIB) and delayed onset and slow buildup of prime mover (GAS) during rise-to-toes task in a subject with PD OFF levodopa medication with improvement on medication compared to a representative healthy control subject. Zero is the onset of peak ankle dorsiflexion. *Adapted from Frank JS, Horak FB, and Nutt JG, Centrally initiated postural adjustments in parkinsonian patients on and off levodopa. Journal of Neurophysiology, 2000. 84(5):2440-8.*

PD showed intact scaling of the magnitude of postural activity. Also, subjects with PD do not perform the rise-to-toes task like normal subjects who are instructed to rise slowly; the relative timing of TIB and GAS activation was different even at comparable speeds of performance.

In addition, analysis of APAs associated with a lateral leg raising task revealed abnormalities.[9,10] In severely affected patients with PD, the amplitude of the initial displacement of the center of foot pressure was markedly lower, and the interval between the earliest force changes and the onset of leg elevation was longer. Furthermore, the alternating bursts and periods of inhibition observed in EMG recordings in normal subjects were replaced by continuous, tonic EMG activity in patients with PD.

APAs should also precede the onset of voluntary movements such as gait initiation.[11,12] In fact, gait initiation involves the correct sequencing of movement preparation and movement execution. The function of APAs is to reduce the effect of the forthcoming body perturbation with anticipatory corrections. In the case of gait initiation, APAs act to accelerate the center of body mass (CoM) forward and laterally toward the stance foot by shifting the center of pressure (CoP) posteriorly and toward the stepping leg. Compared with healthy control subjects, patients with advanced PD generally show APAs abnormally prolonged in duration and reduced in amplitude in both the mediolateral (ML) and anteroposterior (AP) directions (Fig. 5.2). They also show a delay in the sequencing between the beginning of the APA and step onset, accompanied by a shorter and slower first step.

APAs may be impaired in PD because of the many connections of the basal ganglia with the supplementary motor area and the premotor area of the cortex, both of which are implicated in movement preparation[13–15]

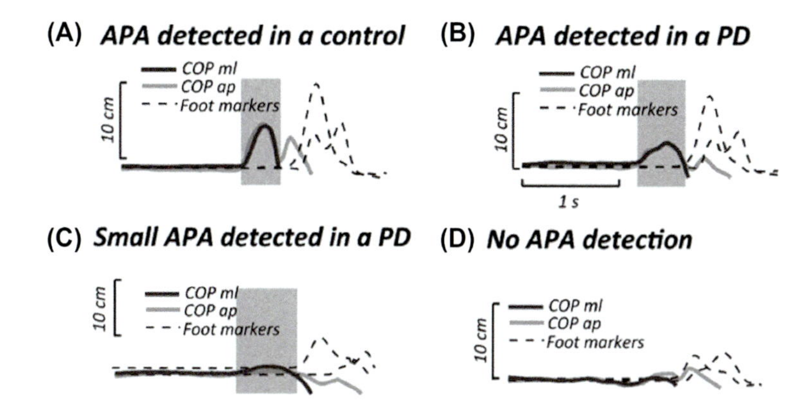

FIGURE 5.2 Examples of APA, as displacement of medial-lateral and anterior posterior CoP, detected prior to start of walking in a healthy control subject (A). Examples of small (B), very small APA (C) and no APA (D) in three subjects with PD are shown for comparison. *Adapted from Validity and reliability of an IMU-based method to detect APAs prior to gait initiation. Mancini M, Chiari L, Holmstrom L, Salarian A, Horak FB. Gait Posture. 2016 Jan;43:125-31.*

or with the peduncular pontine nucleus in the brainstem, which is implicated in locomotion initiation.[16] Gait initiation problems in subjects with PD may also originate from changes in the basal ganglia that result in slowing of the sequential execution of the preparatory and stepping subcomponents of the task.[17]

The prolonged duration of APAs in people with PD may include abnormal pauses that disrupt the posture-movement coordination and may precipitate freezing of gait (FoG). An APA is normally almost always present during voluntary stepping, even in people with FoG. However, it is still controversial whether APAs are smaller or larger than normal associated with freezing (see Chapter 8 for details).

It has been debated whether step initiation is impaired in early stages of PD, before the start of dopaminergic medication. Only a few studies have investigated step initiation in the early phase,[18,19] and only one on a small sample of untreated subjects with PD.[19] A smaller lateral, but not backward, APA magnitude (Fig. 5.3) has been reported in untreated early-to-moderate PD suggesting that the pathology of PD may have a specific effect on loading/unloading of the legs early in the disease, consistent with abnormal ML, but not AP, sway during quiet stance in PD (Chapter 3). Thus, even before just diagnosed patients or clinicians notice "start hesitation" in their patients, APAs can be smaller than normal.

Later in the disease, the magnitude of both lateral and backward APAs becomes bradykinetic in PD.[20] Unlike subjects with early PD, healthy elderly subjects show reduced backward, but not lateral, APA magnitudes compared to young subjects, consistent with separate neural control of these two directions of APAs, which have separate functions.[18] Force for lateral APAs come primarily from hip abductors and ankle extensors, so weakness or poor recruitment of these proximal, axial muscles could

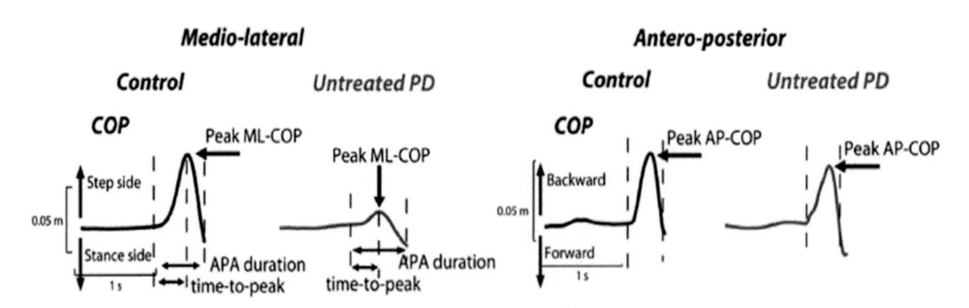

FIGURE 5.3 Even people with early, untreated PD show reduced amplitude of APAs prior to step initiation. Center of pressure (CoP) excursion in the mediolateral (ML) and anteroposterior (AP) direction prior to a step in a representative healthy control and early, untreated subject with PD.[19] *Adapted from Mancini M, Zampieri C, Carlson-Kuhta P, Chiari L, Horak FB. Anticipatory postural adjustments prior to step initiation are hypometric in untreated Parkinson's disease: an accelerometer-based approach. European journal of neurology 2009; 16(9): 1028-34.*

contribute to small lateral APAs in PD. In contrast, force for anterior-posterior APAs come from tibialis anterior muscles, so weakness of ankle dorsiflexors may contribute to small APAs in the elderly.

People with PD also show particularly impaired control of ML sway in quiet stance, consistent with deficits in neural control or proprioception affecting loading and unloading mechanisms with hip muscles that may be important for freezing.[21,22] Since the application of smaller forces to initiate movement results in smaller self-induced postural perturbations, it is also possible that small lateral APAs represent a strategy to minimize postural instability.[18,23]

B. Are APAs context-dependent?

Like all motor programs, the motor program for step initiation needs to change when characteristics of the motor task change. For example, the APA prior to step initiation adapts to initial asymmetric weight loading,[24] and to initial stance on a narrow beam.[25] Studies have shown that PD impairs adaptation of APAs when varying the initial stance width prior to a voluntary step and for reactive postural responses.[26–28]

The preparation for step initiation from wide stance adapts with larger CoP displacements than from narrow stance in elderly and young healthy subjects.[29] Also, the characteristics of the first step (velocity and length) are sensitive to initial stance conditions, probably because of differences in the corresponding APA. In fact, previous studies found a linear correlation between size of lateral CoP displacements during APAs and velocity of locomotion. A larger APA is probably required in wide stance to move the weight off of the initial swing leg, and a larger APA resulted in a longer and quicker step. The changes in step characteristics when the APA changed to accommodate an increased biomechanical demand suggest a close relationship between the program for postural preparation and for stepping.[29]

It is possible that the habitual narrow stance of people with PD is due to their difficulty exerting large APAs from a wider stance in order to shift weight onto one foot. Subjects with moderate to severe PD have much more difficulty initiating a step from a wide stance than from a narrow stance, as shown by the greater differences from control subjects in the magnitude of the APA and by the fact that three PD subjects could not initiate a step at all from wide stance although they could from narrow stance.[29] It is not clear whether PD subjects' increased difficulty in initiating a step from wide stance is due to difficulty in increasing the activation level of muscles for the lateral weight shift due to bradykinesia or due to difficulty scaling or adapting the APA motor program. It is also possible that poor kinesthesia results in inability of motor programs to understand initial postural conditions. Nonetheless, the PD subjects in the

ON levodopa state had the ability to produce larger APAs when switching to a wider stance, suggesting that they do not lack the ability to produce force.

When equilibrium is challenged by an unexpected external displacement of the body, an individual may need to react quickly with an appropriate strategy, that is, stepping to avoid a fall. The execution of a lateral APA before initiating a forward or backward compensatory step, when the body is already accelerating forward, may slow down the most critical action—a step to catch the falling body. Evidence from studies in healthy young and elderly control emphasize that people seldom used an APA in response to fast, unpredictable surface translations.[30,31] It has been hypothesized that the healthy central nervous system down-regulates anticipatory postural preparation when it is not necessary, or, in fact, may delay time to step initiation.[30,31] Surprisingly, evidence shows that forward or backward postural stepping responses to recover from a strong external perturbation may be impaired in people with PD because of excessive postural preparation rather than lack of postural preparation (Fig. 5.4).

Unlike voluntary step initiation, in which people with PD show smaller (or absent) APAs compared with age-matched controls, people with PD show similar size and more APAs prior to compensatory step initiation than age-matched controls.[15,32–34] These findings suggest that compensatory steps by patients with PD are preceded by postural preparation that delays the onset of the step.

It is unclear why people with PD persist in using an APA prior to compensatory stepping. It may be that people with PD and balance deficits cannot appropriately scale down this preplanned anticipatory phase as effectively as control subjects (similarly to the voluntary stepping from narrow and wide stance). Another possibility is that the people with PD compensate for their weak in-place postural responses by slowing down their CoM with a lateral APA to give themselves enough time to take a step with a weak push.

C. Does levodopa improve APAs?

Levodopa markedly improves APAs prior to a voluntary, self-generated movement. Many studies suggest that one essential component of bradykinesia exhibited by PD subjects during voluntary, self-generate movements (either arm raising or step initiation) is an inability to effectively generate muscle activity.[8,29,35–38] Administration of levodopa is associated with increased velocity of upper-extremity voluntary movement,[35,39] rise to toes,[8] and self-generated step initiation, Figs. 5.5.[29,36] Partly in contrast with these findings, only one report showed no effect of levodopa on APAs prior to a voluntary step;[15] however, this result is likely attributable to the

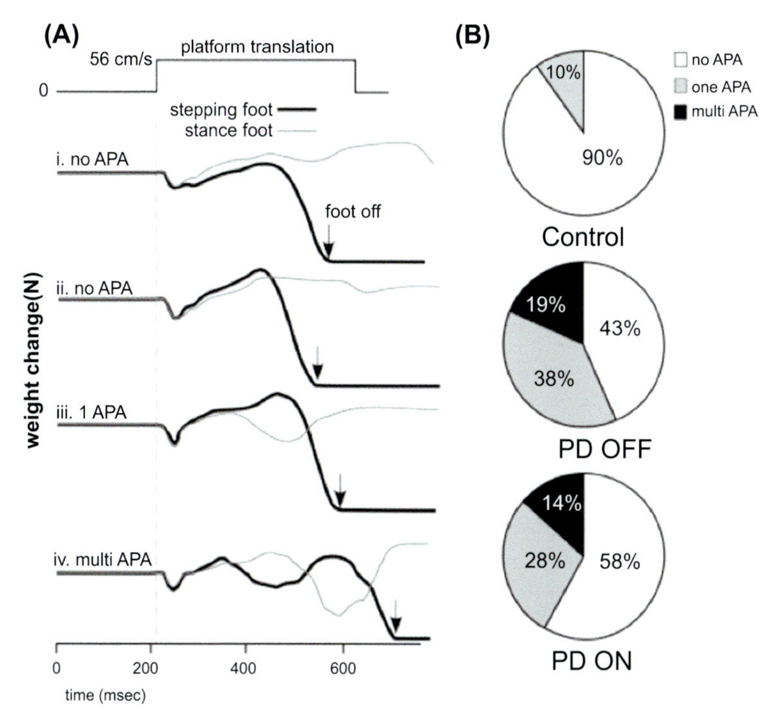

FIGURE 5.4 Different types of postural preparation strategies used in response to surface translation perturbations in Control, PD OFF and PD ON groups.[32] (A). examples of no APA (i and ii), one APA (iii), and multiple APAs (iv) are shown with the weight (newtons) under the stepping foot in black and the weight under the stance foot in gray. (B). The frequency of stepping strategies across the control, PD OFF, and PD ON groups. Out of the percentage of all trials, control subjects rarely exhibited an APA before stepping, whereas the people with PD in both the OFF and ON states commonly used one or more APAs before lifting their foot off the floor. *Adapted from King LA, St George RJ, Carlson-Kuhta P, Nutt JG, Horak FB. Preparation for compensatory forward stepping in Parkinson's disease. Archives of physical medicine and rehabilitation 2010; 91(9): 1332-8.*

participants' foot position since participants were instructed to stand with their feet close together, in which both controls and PD would use a smaller APA.[29] However, in contrast to the efficacy of levodopa for self-generated, voluntary movements, levodopa does not have any effects on APAs generated in preparation for compensatory stepping.[27,32,41]

The selective influence of levodopa therapy on different aspects of postural control suggests that the basal ganglia regulate the different types of postural control by separate neural circuits.[8,42–44] The control of force and sequencing of centrally initiated postural preparations and postural tone appear to involve dopamine circuits because they improve with levodopa therapy.[8,29,36,41] In contrast, the control of force of peripherally triggered postural reactions and APAs prior to compensatory steps, as well as the ability to modify postural synergies based central

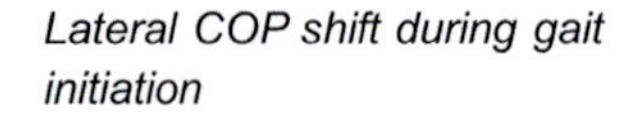

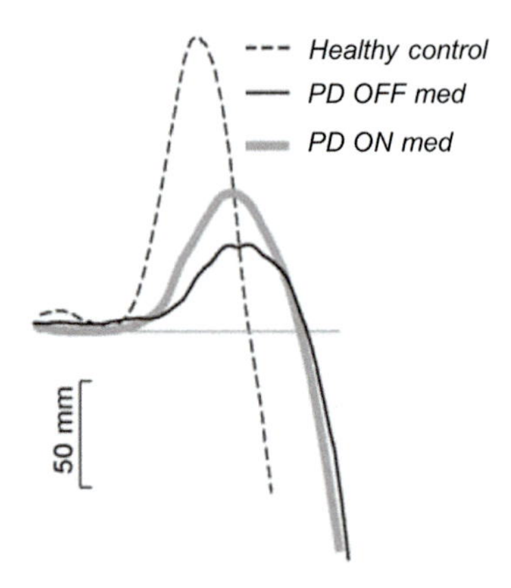

FIGURE 5.5 Improvement of APA amplitude in a PD subject when taking levodopa but still not as large as a healthy control subject's APA. Lateral CoP excursion during APA in a representative healthy control (*dashed line*), subjects with PD OFF (*thin black line*) and ON (*thick gray line*) dopaminergic medication.

set appear to involve nondopaminergic circuits because they are unaffected by levodopa therapy.[27,32,41,45]

Clinicians often observe that speed and force for voluntary movements are improved with levodopa, whereas the ability to resist external perturbations (a push or pull) can be further impaired when ON levodopa. These findings may help to explain why some patients with PD continue to be unstable and fall, although dopamine replacement clearly improves their voluntary movements and speed of walking (see Chapter 6). The failure of levodopa to increase the force of peripherally triggered postural reactions combined with reduction of postural tone with levodopa (Chapter 3) might result in faster falls and greater instability in response to external perturbations.

D. What are the effects of deep brain stimulation on APAs?

The literature on the effects of DBS on APAs is scarce.[40,46–48] Two interesting, small reports[47,48] showed that DBS in the STN produced a marked improvement in APAs' amplitude prior to step initiation. One study examined people with PD from 6 to 31 months after bilateral STN DBS in four conditions: OFF meds/OFF stim, OFF meds/ON stim, ON meds/OFF stim, and ON meds/ON stim.[47] Results show a marked improvement in APAs' amplitude in both ON stim conditions compared to the OFF stim conditions. These findings would suggest that STN DBS is beneficial in improving the preparatory phase of step initiation.[47,48] However, a main flaw of the experimental setup is that it did not allow a comparison of self-generated step initiation prior to, compared to after, surgery. In two other small studies, participants were instructed to start walking in response to a visual triggering signal[48] or a verbal "go" command.[47] Cueing has been found to be very beneficial for gait (Chapter 6) and self-generated movement.[36] See rehabilitation intervention paragraph below. Therefore, these results could indicate that DBS might add a positive effect in improving movements responding to visual or auditory cues.

Our multicenter, double-blinded, clinical trial randomizing STN and GPi DBS[49] also investigated APAs prior to a step in a cohort of 29 subjects with severe PD.[40] Differences in APAs were investigated in two conditions before surgery (OFF/ON levodopa) and in four conditions 6 months after surgery (OFF/ON levodopa combined with OFF/ON DBS). After surgery, the APAs were significantly worse than in the best treatment state before surgery and responsiveness to levodopa decreased,[40] see Fig. 5.6. No differences were detected in effects on APAs between the STN and GPi target groups. In addition, comparison with PD control subjects (who did not receive surgery but were tested 6 m apart) confirmed that deterioration of step preparation was not related to disease progression.

A controversial report on seven subjects who underwent bilateral STN DBS with one contact of each electrode also located deeper within the substantia nigra pars reticulata (SNr) show some evidence that different

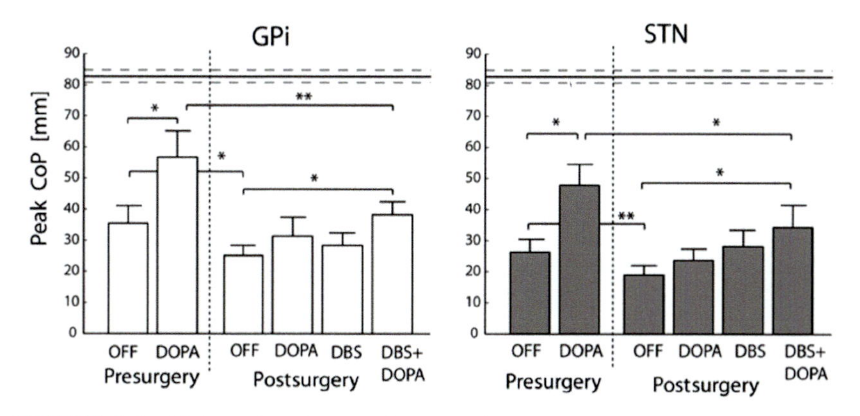

FIGURE 5.6 DBS in either GPi or in STN significantly decreases the peak amplitude of APAs prior to step initiation compared to prior to surgery 6 months earlier $^{**}P < .01$). Levodopa significantly increases APA amplitude prior to DBS surgery but not following DBS surgery. Bar graph representing the lateral CoP excursion for the APA before and after surgery in 15 people with GPi and 14 with STN DBS stimulation compared to a group of 10 elderly control subjects (mean and SE in horizontal lines).[40] *Adapted from Rocchi L, Carlson-Kuhta P, Chiari L, Burchiel KJ, Hogarth P, Horak FB. Effects of deep brain stimulation in the subthalamic nucleus or globus pallidus internus on step initiation in Parkinson disease: laboratory investigation. Journal of neurosurgery 2012; 117(6): 1141-9.*

brain areas are responsible for different aspects of locomotion.[46] The effects of these treatments on motor parkinsonian disability were assessed with the UPDRS III scale, separated into "axial" (rising from chair, posture, postural stability, and gait) and "distal" scores. Whereas levodopa and/or STN stimulation improved "axial" and "distal" motor symptoms and step execution, SNr stimulation improved only the "axial" symptoms and braking capacity. Specifically, the findings from Chastan et al.[46] suggest that anteroposterior (length and velocity) and vertical (braking capacity) parameters for step initiation are controlled by two distinct systems within the basal ganglia circuitry, representing, respectively, locomotion and balance.[46] The SNr, a major basal ganglia output known to project to pontomesencephalic structures, may be particularly involved in balance control during gait.[46] However, these results have yet to be confirmed in a larger sample size on clearer measures of preparation and execution of self-generate step initiation.

E. Can rehabilitation affect APAs?

External sensory cues (auditory, visual, and vibrotactile) have been shown to improve self-generated step initiation in subjects with PD.[15,36,50–53] Despite an overall effectiveness in improving step initiation characteristics almost to normative values, Lu et al.[53] demonstrated that cueing effects on APA timing varied across modalities. Specifically, visual and acoustic cues reduced APA and toe-off timings while vibrotactile cues had no significant effect on APA timing and had reduced efficacy on APA amplitude compared to the acoustic and visual cues. The reduced effectiveness of vibrotactile stimulation found in Lu partly contradicts other results.[51] A possible explanation is the difference in stimulation location, in fact, while stimulation on the lateral malleolus of the initial stance limb would reinforce the APA sequence[54-56] often, people with PD show compromised proprioceptive sensation. The positive effect of external cueing may be partially explained as external timing cue used for enhancing the sequencing of motor subtasks that comprise a complex plan of action.

Cueing might be an effective rehabilitative strategy to improve preparation of gait initiation, independently of the type of cue. However, further research is necessary to understand which type of cue best improves the preparation and execution of gait initiation, to explain which mechanisms are involved during cueing, and to establish retention. In fact, the mechanisms by which external cues improve APAs have yet to be described; brainstem networks[43] and/or executive[57] and attentional networks[58] may be involved.

Findings on immediate and short-term effects of externally applied robotic assistance[56,59,60] for APAs in PD show positive results. While subjects with PD have generally longer APA and step duration than control subjects, the use of the lateral assist for gait initiation improved performance such that PD subjects' APA duration resembled the baseline performance of the healthy controls, while their step duration was reduced.[56,59,60] The presentation of a precisely controlled lateral waist-pull introduced shortly after the self-initiated onset of the mediolateral APA assisted the lateral weight transfer by producing a faster APA duration. This postural effect was followed by an earlier step initiation onset time and faster execution time for PD subjects. These findings indicate a capacity in patients with early stage PD to improve their step timing characteristics by affecting posture and locomotion interactions through the use of anticipatory postural assistance.

The lateral assist could enable the prerequisite postural state of change from bipedal to single limb support that may be anticipated by the CNS before the release of the step cycle is initiated.[23] In this form of motor prediction, the neural circuits for initiating stepping could be actively delayed until APAs have reached a state condition associated with lateral weight transfer and postural stability.

Lastly, the effects of a progressive resistance training (PRT) on step initiation has been investigated in a randomized study in 18 subjects with PD[61]. PRT may be a particularly useful intervention in PD where reduced muscle strength is associated with disease related reductions in central activation and reduced physical activity. Results show an improvement in APA amplitude (larger APA) and first step execution (shorter time) only in the PRT group, and not in the control group. These findings might result from a greater propulsive force production by the swing limb and greater ability of the hip flexors to swing the limb forward after training.[61] PRT may also enhance control of proprioception.

The fact that emotion, attention, external triggers, and dopaminergic drugs can all modify the APA motor program suggests the existence of a complex pathophysiological mechanism that involves not only locomotor networks but also cortical areas and the basal ganglia system.[50] Abnormal coupling between standing posture and APAs and between APAs and step execution appears to be a crucial part of the pathophysiological mechanism. Although external cueing appears to be helpful in triggering larger APAs, only a few studies have provided evidence of the efficacy of various rehabilitation methods in routine care.

Highlights

- APAs are slower and hypometric in PD from early in the disease.
- APAs are inflexible in PD; in fact, people with PD show an impaired ability to scale their APAs for step initiation with a wider stance compared to a narrow stance. Also, people with PD use APAs prior to protective steps, whereas controls do not.
- Levodopa improves APAs, although not to control values.
- DBS findings are not yet conclusive; however, it appears that both GPi and STN DBS are detrimental for APAs.
- Rehabilitation, especially in the form of external cueing or assisted transitions, seems to be beneficial for APAs, although the best type of rehabilitation or long-term effects on APAs in people with PD is unknown.

References

1. Bleuse S, Cassim F, Blatt JL, et al. Anticipatory postural adjustments associated with arm movement in Parkinson's disease: a biomechanical analysis. *Journal of Neurology Neurosurgery and Psychiatry* 2008;**79**(8):881–7.
2. Cordo PJ, Nashner LM. Properties of postural adjustments associated with rapid arm movements. *Journal of Neurophysiology* 1982;**47**(2):287–302.
3. Bazalgette D, Zattara M, Bathien N, Bouisset S, Rondot P. Postural adjustments associated with rapid voluntary arm movements in patients with Parkinson's disease. *Advances in Neurology* 1987;**45**:371–4.
4. Bouisset S, Zattara M. Biomechanical study of the programming of anticipatory postural adjustments associated with voluntary movement. *Journal of Biomechanics* 1987;**20**(8):735–42.
5. Dick JP, Rothwell JC, Berardelli A, et al. Associated postural adjustments in Parkinson's disease. *Journal of Neurology Neurosurgery and Psychiatry* 1986;**49**(12):1378–85.
6. Rogers MW, Kukulka CG, Soderberg GL. Postural adjustments preceding rapid arm movements in parkinsonian subjects. *Neuroscience Letters* 1987;**75**(2):246–51.
7. Diener HC, Dichgans J, Guschlbauer B, Bacher M, Rapp H, Langenbach P. Associated postural adjustments with body movement in normal subjects and patients with parkinsonism and cerebellar disease. *Revue Neurologique* 1990;**146**(10):555–63.
8. Frank JS, Horak FB, Nutt J. Centrally initiated postural adjustments in parkinsonian patients on and off levodopa. *Journal of Neurophysiology* 2000;**84**(5):2440–8.
9. Defebvre LJ, Krystkowiak P, Blatt JL, et al. Influence of pallidal stimulation and levodopa on gait and preparatory postural adjustments in Parkinson's disease. *Movement Disorders* 2002;**17**(1):76–83.
10. Lee RG, Tonolli I, Viallet F, Aurenty R, Massion J. Preparatory postural adjustments in parkinsonian patients with postural instability. *The Canadian Journal of Neurological Sciences* 1995;**22**(2):126–35.
11. Bouisset S, Do MC. Posture, dynamic stability, and voluntary movement. *Neurophysiologie Clinique* 2008;**38**(6):345–62.
12. Breniere Y, Do MC. When and how does steady state gait movement induced from upright posture begin? *Journal of Biomechanics* 1986;**19**(12):1035–40.

13. Massion J. Movement, posture and equilibrium: interaction and coordination. *Progress in Neurobiology* 1992;**38**(1):35–56.
14. Nakano K, Kayahara T, Tsutsumi T, Ushiro H. Neural circuits and functional organization of the striatum. *Journal of Neurology* 2000;**247**(Suppl. 5):V1–15.
15. Schlenstedt C, Mancini M, Horak F, Peterson D. Anticipatory postural adjustment during self-initiated, cued, and compensatory stepping in healthy older adults and patients with Parkinson disease. *Archives of Physical Medicine and Rehabilitation* 2017;**98**(7): 1316–24. e1.
16. Pahapill PA, Lozano AM. The pedunculopontine nucleus and Parkinson's disease. *Brain* 2000;**123**(Pt 9):1767–83.
17. Rosin R, Topka H, Dichgans J. Gait initiation in Parkinson's disease. *Movement Disorders* 1997;**12**(5):682–90.
18. Carpinella I, Crenna P, Calabrese E, et al. Locomotor function in the early stage of Parkinson's disease. *IEEE Transactions on Neural Systems and Rehabilitation Engineering* 2007; **15**(4):543–51.
19. Mancini M, Zampieri C, Carlson-Kuhta P, Chiari L, Horak FB. Anticipatory postural adjustments prior to step initiation are hypometric in untreated Parkinson's disease: an accelerometer-based approach. *European Journal of Neurology* 2009;**16**(9):1028–34.
20. Halliday SE, Winter DA, Frank JS, Patla AE, Prince F. The initiation of gait in young, elderly, and Parkinson's disease subjects. *Gait & Posture* 1998;**8**(1):8–14.
21. Dietz V, Duysens J. Significance of load receptor input during locomotion: a review. *Gait & Posture* 2000;**11**(2):102–10.
22. Vaugoyeau M, Viel S, Assaiante C, Amblard B, Azulay JP. Impaired vertical postural control and proprioceptive integration deficits in Parkinson's disease. *Neuroscience* 2007;**146**(2):852–63.
23. Vaugoyeau M, Viallet F, Mesure S, Massion J. Coordination of axial rotation and step execution: deficits in Parkinson's disease. *Gait & Posture* 2003;**18**(3):150–7.
24. Patchay S, Gahery Y. Effect of asymmetrical limb loading on early postural adjustments associated with gait initiation in young healthy adults. *Gait & Posture* 2003;**18**(1):85–94.
25. Couillandre A, Breniere Y, Maton B. Is human gait initiation program affected by a reduction of the postural basis? *Neuroscience Letters* 2000;**285**(2):150–4.
26. Chong RK, Horak FB, Woollacott MH. Parkinson's disease impairs the ability to change set quickly. *Journal of the Neurological Sciences* 2000;**175**(1):57–70.
27. Horak FB, Nutt JG, Nashner LM. Postural inflexibility in parkinsonian subjects. *Journal of the Neurological Sciences* 1992;**111**(1):46–58.
28. Schieppati M, Nardone A. Free and supported stance in Parkinson's disease. The effect of posture and 'postural set' on leg muscle responses to perturbation, and its relation to the severity of the disease. *Brain* 1991;**114**(Pt 3):1227–44.
29. Rocchi L, Chiari L, Mancini M, Carlson-Kuhta P, Gross A, Horak FB. Step initiation in Parkinson's disease: influence of initial stance conditions. *Neuroscience Letters* 2006; **406**(1–2):128–32.
30. Henry SM, Fung J, Horak FB. Control of stance during lateral and anterior/posterior surface translations. *IEEE Transactions on Rehabilitation Engineering* 1998;**6**(1):32–42.
31. McIlroy WE, Maki BE. Do anticipatory postural adjustments precede compensatory stepping reactions evoked by perturbation? *Neuroscience Letters* 1993;**164**(1–2):199–202.
32. King LA, St George RJ, Carlson-Kuhta P, Nutt JG, Horak FB. Preparation for compensatory forward stepping in Parkinson's disease. *Archives of Physical Medicine and Rehabilitation* 2010;**91**(9):1332–8.
33. Peterson DS, Lohse KR, Mancini M. Anticipatory postural responses prior to protective steps are not different in people with PD who do and do not freeze. *Gait & Posture* 2018; **64**:126–9.

34. Peterson DS, Lohse KR, Mancini M. Relating anticipatory postural adjustments to step outcomes during loss of balance in people with Parkinson's disease. *Neurorehabilitation and Neural Repair* 2018;**32**(10):887–98.
35. Berardelli A, Accornero N, Argenta M, Meco G, Manfredi M. Fast complex arm movements in Parkinson's disease. *Journal of Neurology Neurosurgery and Psychiatry* 1986;**49**(10):1146–9.
36. Burleigh-Jacobs A, Horak FB, Nutt JG, Obeso JA. Step initiation in Parkinson's disease: influence of levodopa and external sensory triggers. *Movement Disorders* 1997;**12**(2):206–15.
37. Godaux E, Koulischer D, Jacquy J. Parkinsonian bradykinesia is due to depression in the rate of rise of muscle activity. *Annals of Neurology* 1992;**31**(1):93–100.
38. Hallett M, Khoshbin S. A physiological mechanism of bradykinesia. *Brain* 1980;**103**(2):301–14.
39. Johnson MT, Mendez A, Kipnis AN, Silverstein P, Zwiebel F, Ebner TJ. Acute effects of levodopa on wrist movement in Parkinson's disease. Kinematics, volitional EMG modulation and reflex amplitude modulation. *Brain* 1994;**117**(Pt 6):1409–22.
40. Rocchi L, Carlson-Kuhta P, Chiari L, Burchiel KJ, Hogarth P, Horak FB. Effects of deep brain stimulation in the subthalamic nucleus or globus pallidus internus on step initiation in Parkinson disease: laboratory investigation. *Journal of Neurosurgery* 2012;**117**(6):1141–9.
41. Horak FB, Frank J, Nutt J. Effects of dopamine on postural control in parkinsonian subjects: scaling, set, and tone. *Journal of Neurophysiology* 1996;**75**(6):2380–96.
42. Mackinnon C. Sensorimotor anatomy of gait, balance, and falls. In: *Handbook of Clinical Neurology*. Elsevier; 2018.
43. MacKinnon CD, Bissig D, Chiusano J, et al. Preparation of anticipatory postural adjustments prior to stepping. *Journal of Neurophysiology* 2007;**97**(6):4368–79.
44. Rogers MW, Kennedy R, Palmer S, et al. Postural preparation prior to stepping in patients with Parkinson's disease. *Journal of Neurophysiology* 2011;**106**(2):915–24.
45. Peterson DS, Horak FB. The effect of levodopa on improvements in protective stepping in people with Parkinson's disease. *Neurorehabilitation and Neural Repair* 2016;**30**(10):931–40.
46. Chastan N, Westby GW, Yelnik J, et al. Effects of nigral stimulation on locomotion and postural stability in patients with Parkinson's disease. *Brain* 2009;**132**(Pt 1):172–84.
47. Crenna P, Carpinella I, Rabuffetti M, et al. Impact of subthalamic nucleus stimulation on the initiation of gait in Parkinson's disease. *Experimental Brain Research* 2006;**172**(4):519–32.
48. Liu W, McIntire K, Kim SH, et al. Bilateral subthalamic stimulation improves gait initiation in patients with Parkinson's disease. *Gait & Posture* 2006;**23**(4):492–8.
49. Weaver FM, Follett K, Stern M, et al. Bilateral deep brain stimulation vs best medical therapy for patients with advanced Parkinson disease: a randomized controlled trial. *Jama* 2009;**301**(1):63–73.
50. Delval A, Tard C, Defebvre L. Why we should study gait initiation in Parkinson's disease. *Neurophysiologie Clinique* 2014;**44**(1):69–76.
51. Dibble LE, Nicholson DE, Shultz B, MacWilliams BA, Marcus RL, Moncur C. Sensory cueing effects on maximal speed gait initiation in persons with Parkinson's disease and healthy elders. *Gait & Posture* 2004;**19**(3):215–25.
52. Gantchev N, Viallet F, Aurenty R, Massion J. Impairment of posturo-kinetic coordination during initiation of forward oriented stepping movements in parkinsonian patients. *Electroencephalography and Clinical Neurophysiology* 1996;**101**(2):110–20.
53. Lu C, Amundsen Huffmaster SL, Tuite PJ, Vachon JM, MacKinnon CD. Effect of cue timing and modality on gait initiation in Parkinson disease with freezing of gait. *Archives of Physical Medicine and Rehabilitation* 2017;**98**(7). 1291-9.e1.

54. Kukulka CG, Hajela N, Olson E, Peters A, Podratz K, Quade C. Visual and cutaneous triggering of rapid step initiation. *Experimental Brain Research* 2009;**192**(2):167–73.
55. Creath RA, Prettyman M, Shulman L, et al. Self-triggered assistive stimulus training improves step initiation in persons with Parkinson's disease. *Journal of Neuroengineering and Rehabilitation* 2013;**10**:11.
56. Mille ML, Creath RA, Prettyman MG, et al. Posture and locomotion coupling: a target for rehabilitation interventions in persons with Parkinson's disease. *Parkinson's Disease* 2012;**2012**:754186.
57. Rochester L, Baker K, Hetherington V, et al. Evidence for motor learning in Parkinson's disease: acquisition, automaticity and retention of cued gait performance after training with external rhythmical cues. *Brain Research* 2010;**1319**:103–11.
58. Azulay JP, Mesure S, Blin O. Influence of visual cues on gait in Parkinson's disease: contribution to attention or sensory dependence? *Journal of the Neurological Sciences* 2006;**248**(1–2):192–5.
59. Mille ML, Hilliard MJ, Martinez KM, Simuni T, Zhang Y, Rogers MW. Short-term effects of posture-assisted step training on rapid step initiation in Parkinson's disease. *Journal of Neurologic Physical Therapy* 2009;**33**(2):88–95.
60. Rogers MW, Hilliard MJ, Martinez KM, Zhang Y, Simuni T, Mille ML. Perturbations of ground support alter posture and locomotion coupling during step initiation in Parkinson's disease. *Experimental Brain Research* 2011;**208**(4):557–67.
61. Hass CJ, Buckley TA, Pitsikoulis C, Barthelemy EJ. Progressive resistance training improves gait initiation in individuals with Parkinson's disease. *Gait & Posture* 2012;**35**(4):669–73.

CHAPTER

6

How is dynamic balance during walking affected by PD?

Clinical case

Betty first noticed something was wrong with her walking when her husband noted that she did not swing one arm as much as the other, and she had trouble lifting one foot high enough to avoid a trip when she ran. After she was diagnosed with Parkinsonian disease (PD), her testing showed that she turned more slowly and took more steps to turn than normal even though her gait speed was still in the normal range for her age. As the disease progressed, she walked with short, shuffling steps and often fell as she tripped over obstacles. Even though levodopa greatly increased her stride length and walking speed, she continued to fall and became increasingly fearful of walking. Later, deep brain stimulation (DBS) further improved her speed of walking, but she found herself falling even more than before the surgery. Luckily, Betty found an experienced physical therapist followed by a sustained community balance exercise program that reduced her falls while maintaining her mobility.

A. How does impaired balance affect gait and impaired gait affect balance in PD?

Over 50% of all falls in people with PD occur during walking.[1,2] PD is associated with more unstable gait during unperturbed walking compared to elderly controls, Fig. 6.1 represents the classic features of walking in PD, like flexed trunk, lack of arm swing, and narrow and short steps. Gait in people with PD is also characterized by increased temporal variability and impaired anticipatory and reactive balance control during walking.

https://doi.org/10.1016/B978-0-12-813874-8.00006-4

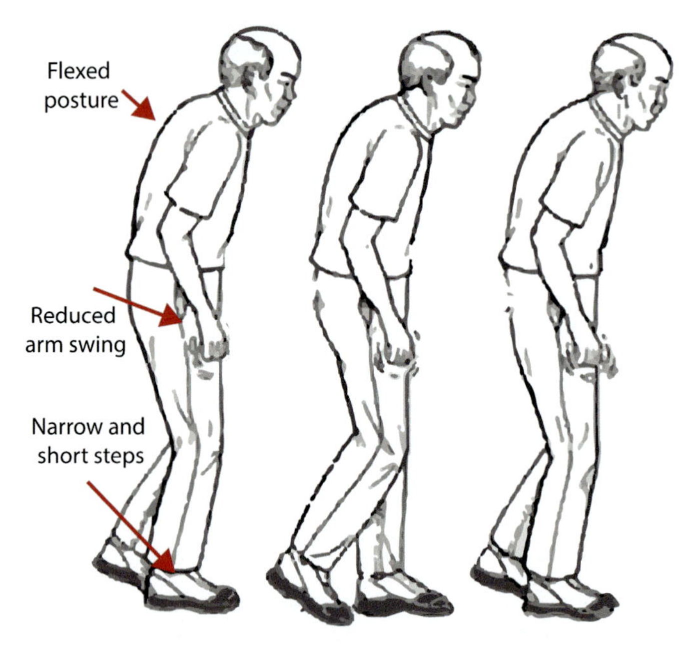

FIGURE 6.1 Characteristic features of parkinsonian gait. *Slowness*—narrow strides, short shuffling steps, reduced upper body motion, and slow cadence; *Posture*—flexed and asymmetrical postural alignment; *Variability/asymmetry*—variable, asymmetrical step timing and length; *Upper body*—reduced arm swing and trunk motion; and *Balance*—weak postural responses during walking and slow, jerky turning.

During walking, people with PD show ~40% lower margin of stability than age-matched elderly people, based on the extrapolated center of mass over the base of foot support.[3] The extrapolated center of mass includes both the position and velocity of the body center of mass relative to the limits of stability. The lower margin of stability during walking is due to a smaller base of foot support (narrow stance width, short steps, and flat-footed shuffle) and a more anterior position of body center of mass related to a flexed posture. This smaller limits of stability during walking makes it more difficult to prospectively control the upper body over the foot support and more difficult to respond to unexpected gait perturbations.[3]

The ability to walk quickly and safely in a variety of environments severely affects perceived quality of life for people with PD.[2] Walking impairments also closely track with severity of disease. For example, severity of PD, as measured with the Motor MDS-UPDRS III scale (including limb bradykinesia, tremor, rigidity, as well as four postural instability and gait disability (PIGD) items related to postural alignment, sit-to-stand, walking, and response to a backward shoulder pull, respectively) is often associated with pace characteristics of walking, such as gait velocity.[4] However, the specific PD gait impairment best related to the PIGD items of the Motor UPDRS may be dynamic balance, or control of the trunk while walking.

To determine how many different types of neural control factors are involved in gait, studies have used factor analysis or principle component analysis to determine the amount of variance accounted for by groups of measures that are related to each other. For example, a factor analysis of 20 gait and 12 postural sway measures in 100 moderately affected people with PD, Off (and then On) their medication, showed that the gait and sway measures were independent (Fig. 6.2). In addition, gait was decomposed into four different factors; speed (or pace), upper body arm/

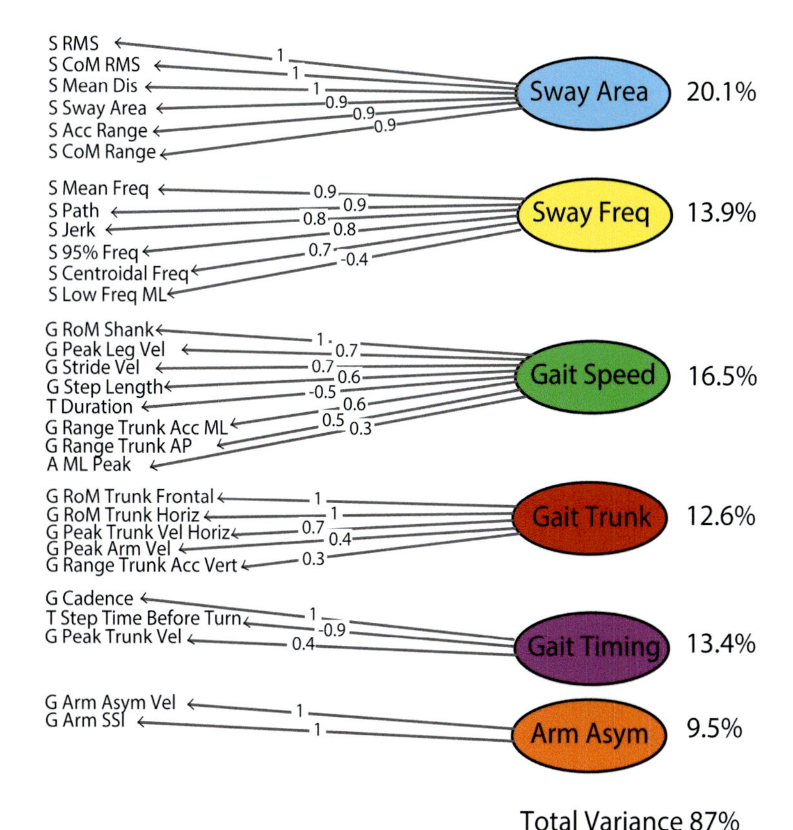

FIGURE 6.2 Factor analysis of 30 measures performed during an Instrumented Stand and Walk test in 100 people with PD in the Off state. Six independent domains of gait and balance were identified: Sway Area, Sway Frequency, Gait Speed, Gait Trunk, Gait Timing, and Arm Asymmetry. The percent of variance explained by each factor and the total variance is given at right, as well as the loading of each measure. *Adapted from Horak FB, Mancini M, Carlson-Kuhta P, Nutt JG, Salarian A. Balance and gait represent independent domains of mobility in Parkinson disease.* Physical Therapy *2016;**96**(9):1364–71.*

trunk, timing/turning, and arm asymmetry (Fig. 6.2). Six factors (sway area, sway frequency, gait speed (pace), trunk control, gait timing, and arm asymmetry) explain 87% of the variance among the cohort. The fact that only the speed (pace) of gait significantly improved with levodopa supports the notion that a variety of different brain networks control gait.

Fig. 6.3 shows the significant relationship between the PIGD subscore of the UPDRS and trunk metrics (a factor based on 5 gait metrics from Fig. 6.3).[5] However, the other gait factors in Fig. 6.2 did not significantly correlate with the PIGD, emphasizing the importance of measuring axial balance control for functional dynamic balance during walking.[5] In this study, levodopa improved only gait speed and worsened sway area without significantly changing dynamic balance or timing measures of gait. This change in some, but not all, factors characterizing gait disorders in PD emphasizes the relatively independent control of different aspects of gait by the nervous system.

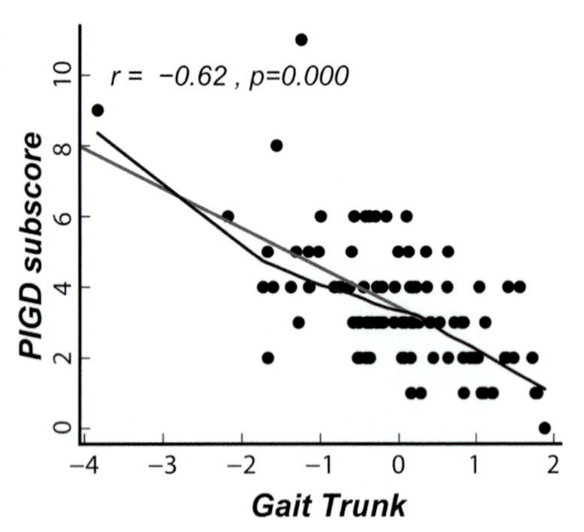

FIGURE 6.3 Linear model relating the postural instability and gait disability (PIGD) subscore (from UPDRS Part III) with a set of gait metrics in the Gait Trunk factor (Figure 6.2).

Other studies using factor analysis or principal component analysis, in either subjects with PD On medication or healthy elderly subjects are essentially in agreement; gait pace, variability, asymmetry, and balance factors emerge as independent factors from foot-fall-based gait measures that did not include postural sway or upper body measures.[6–8]

However, it is hard to determine which specific characteristics of parkinsonian gait are responsible for falls, which characteristics compensate for balance impairments to reduce falls, and which gait characteristics are not related to falls. The abnormal walking pattern in PD is characterized by short shuffling steps, slower gait speed, and longer double-support phase than those of healthy control subjects.[9–12] PD gait also is characterized by reduced rotation of the trunk and pelvis,[13] higher stride-to-stride variability[9,14–16] and turning involves more, and smaller, steps.[13,17–21] It is unclear whether these changes to gait are attempts to prevent falls due to impaired control of balance and which changes are primary to PD signs, such as bradykinesia, and thus, result in worse balance control.

Studies agree that although balance and gait are both affected by PD and both are critical for functional mobility, balance and gait represent independent, albeit interacting, functional neural control systems. That is, parkinsonian gait characteristics can impair balance control and parkinsonian balance disorders can impair gait control, and it is often difficult to determine which impaired gait or balance metrics are primary or compensatory characteristics.

Additional evidence that balance and gait are independent comes from studies in newly diagnosed people with PD showing that balance problems precede gait problems. Postural sway in standing is abnormal in untreated, newly diagnosed people with PD (Chapter 3), even when their gait speed is still in the normal range.[22] In addition to abnormal postural sway, turning velocity (Chapter 7) and arm swing velocity are also slow, long before straight ahead gait speed has slowed.[23,24] Turning is a dynamic balance task unlike straight ahead walking (see Chapter 7). Arm swing is associated with upper trunk control, critical for balance control while walking.

Turning velocity, arm swing velocity, and arm swing asymmetry are very early signs of gait disturbance in newly diagnosed people with PD.[23] Our pilot study suggests that turning and arm swing measures are also the gait measures most likely to show progression over the first 18 months

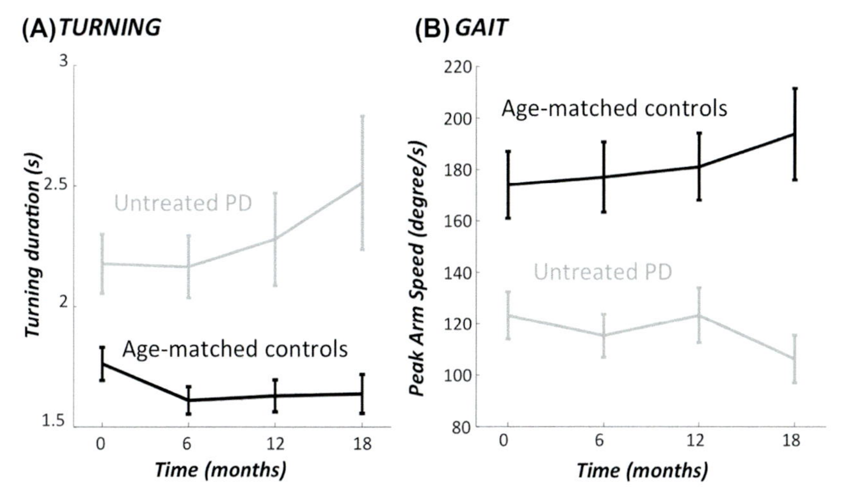

FIGURE 6.4 A. Turning duration for 180 degrees turns while walking and B. arm swing speed during walking shows progression in newly diagnosed people with PD but not in age-matched control subjects.

of early, untreated PD (unpublished results in Figure 6.4). Twelve people with untreated mild-to-moderate idiopathic PD and 12 age-matched control subjects performed an instrumented Get Up and Go test with wearable inertial sensors[23] every 6 months for 18 months. From all of the gait and standing postural sway measures in Fig. 6.2, turning and arm speed were the most sensitive to early disease progression.

In fact, arm swing asymmetry and gait variability, another sign of abnormal dynamic balance,[9] are very sensitive to prodromal PD, in people who will eventually manifest PD with clinically observable parkinsonian signs because of the LRRK2-G2019S mutation.[24] A large study on a total of 380 participants (186 healthy nonmanifesting controls and 196 participants with PD) showed that nonmanifesting participants who were carriers of the G2019S mutation walked with greater arm swing asymmetry and variability and smaller axial trunk rotation smoothness when compared with noncarriers. In the nonmanifesting mutation carriers, arm swing asymmetry was also associated with gait stride time variability.[24]

Since impairments of gait and balance have been shown to be relatively independent,[2,5] we postulate that gait and balance control depend on distinct neural networks.[2] In addition to continuous characteristics of gait that affect balance, transient gait disturbances such as occasional, context-specific festination (rapid, short steps) or freezing of gait (either akinesia or trembling of the legs) are often associated with falling[25–27] (see also Chapter 8).

Control of balance while walking depends upon using a functional locomotor pattern that allows controlled forward progression of the body center of mass (CoM) in the head/arms/trunk (HAT) segment (minimizing lateral HAT motion) over the reciprocal, swing and stance phases of the gait cycle that must anticipate and prevent the falling motion of the body center of mass. Control of balance while walking also depends upon multisensory control of upright trunk orientation and equilibrium (Chapter 3), automatic postural responses to slips and trips (Chapter 4), as well as anticipatory postural adjustments that shift the body center of mass over the stance leg to swing the unweighted leg forward (Chapter 5).

Slowness

It is not clear whether the most easily recognizable characteristic of gait in people with PD, slowness, is due primarily to bradykinesia or to their balance disorders. Indeed, fear of falling, impaired control of upper body displacement, and reduced size/force of reactive and anticipatory postural control are associated with slower comfortable gait speed in the elderly.[28,29] The slow gait in people with PD is due not only to bradykinesia, but also to muscle weakness. Bradykinesia and muscle weakness share the same underlying physiological mechanisms in PD[30]; that is, insufficient motor unit recruitment as a result of reduced excitatory drive to the motor cortex caused by dopaminergic deficit. Abnormal muscle activation patterns in PD consist of reduced intensity of agonist bursts, excessive antagonist activation, altered muscle activation frequencies, and reduced rate of torque generation. Muscle weakness in PD is central in origin, so it can be improved with the administration of antiparkinson medications.[31] The reduced lower-limb strength in PD is associated with postural instability, poorer physical functioning and walking performance, and increased risk of falls.[32]

Slowness of gait caused by bradykinesia and weakness may compromise postural stability during walking. Once walking has been successfully initiated with an anticipatory postural adjustment (Chapter 5), people with PD may find themselves vulnerable to a fall for many reasons related to the parkinsonian signs of bradykinesia and rigidity that result in weak, push-off forces during gait. Although elderly people with balance disorders slow their gait to compensate for poor balance control, slowness of gait in PD is likely also due to their primary neurological motor signs of bradykinesia and rigidity, as well as to their fatigability and secondary sarcopenia or muscle weakness.

Slowness can cause falls for many reasons, such as slow weight shifting associated with inability to clear the floor with the swing leg. Slowness of walking is also associated with narrow strides, short and shuffling steps vulnerable to tripping, and reduced upper/lower body counterrotations that increase walking energetics. Slowness compromises balance because slow cadence results in long double-support time, perhaps in an attempt to compensate for single foot balance. In addition to slowness, the flexed posture of parkinsonism may result in the body CoM falling forward ahead of the base of foot support (propulsion) with failure take a large enough step to halt the fall of the CoM. Rigidity is associated with muscle cocontraction that could make it difficult to compensate for variable lateral trunk instability with varying foot support time or position.[33] Alternatively, some people with PD fall because force generation for postural responses is too slow to recover from a slip or trip. They may also trip more often than normal because they fail to lift their foot high enough over obstacles.[34]

Slowness of gait is the impairment most related to self-perceived mobility. Fig. 6.5 is a spider plot to compare the size of correlations between individual gait metrics and the patient-reported outcomes of (A) Activities of Balance Confidence Scale (ABC) and (B) the Parkinson Disease Quality of Life Mobility subcomponent (PDQ) in 100 people with PD in the On and Off levodopa states.[4] Results showed that stride velocity, stride length, leg range of motion, and turning speed were most highly correlated to patient perception of mobility disability. These same

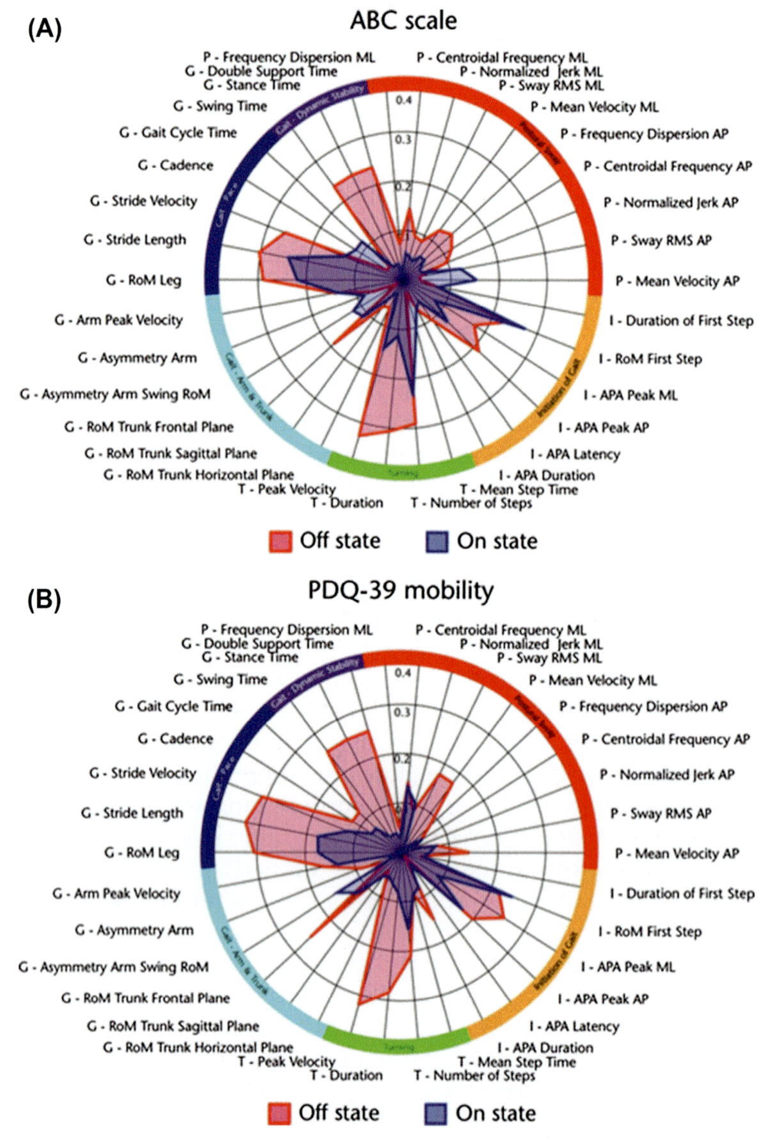

FIGURE 6.5 Radar plot comparing Spearman correlation (absolute values) of (A) Activities-specific Balance Confidence scale (ABC scale) and (B) Mobility domain of the Parkinson Disease Questionnaire (PDQ-39 mobility items 1–10) with measures of gait and balance function in the OFF- and ON-medication states. Adapted from Curtze C, Nutt JG, Carlson-Kuhta P, Mancini M, Horak FB. Objective Gait and Balance Impairments Relate to Balance Confidence and Perceived Mobility in People With Parkinson Disease. Physical therapy 2016; 96(11): 1734-43.

measures were also most strongly related to severity of disease using the UPDRS Motor Score.[4] Surprisingly, the objective measures of gait measured in the OFF-medication state were more indicative of patient perception of balance confidence and mobility disability than the ON-medication state gait measures.[4] Thus, people with PD must be thinking back to the moments during their daily life that they are most affected when their medication is not working optimally.

Gait variability/asymmetry

Gait variability reflects stride-to-stride fluctuations in walking[35] and is consistently larger in people with PD compared to age-matched control subjects.[9,36] However, different types of gait variability may represent different types of neural control/dyscontrol. Lateral step variability and lateral trunk motion variability may reflect dynamic postural control, whereas, anterior-posterior gait variability (such as stride time and stride length variability) is related to natural fluctuations in self-selected walking speed over time. Balance control while walking involves compensating for lateral tilt of the trunk and body CoM with subtle adjustments of lateral foot placement and duration of foot stance time.[33,37] Studies and models have demonstrated how step width variability in the lateral direction reflect active step-by-step adjustments by the nervous system to maintain balance during walking.[38] In contrast, anterior-posterior stride time and stride length variability are higher the slower people walk, and so reflect natural fluctuations in walking speed.[39] Interestingly, unlike stride time variability, swing time variability is unrelated to walking speed so may be used as a speed-independent marker of rhythmicity and gait steadiness and suggest that increased gait variability in PD is disease-related, and not simply a consequence of bradykinesia.[39] However, a reliable measure of gait variability requires a long walk with at least 30 steps.[40]

Thus, some types of increased gait variability in PD are disease related and not simply a consequence of bradykinesia. In fact, gait variability is a more sensitive predictor of falls than gait speed[41] and discriminates, better than gait speed, presymptomatic carriers of the LRRK2-G2019S mutation as a precursor of PD compared to noncarriers.[24] Gait variability during daily life showed that very different types of variability distinguished PD fallers from nonfallers.[41] Whereas PD fallers walked with shorter and less variable ambulatory bouts than nonfallers, fallers walked with more variable step length than nonfallers.[41] Prodromal PD in carriers of LRRK2 showed higher intraindividual variability in arm motion activity during walking.[42]

It has been hypothesized that increased variability of gait in PD reflects less automatic control of walking.[43] The theory is that loss of basal ganglia subcortical control of locomotion would result in a compensatory shift to cortical or cerebellar control of locomotion, which may result in excessive gait variability.[44,45] In fact, impairments in motor automaticity are a hallmark feature of PD that cause patients to increasingly demand cortical resources in order to execute basic motor operations via attentional processes.[46] A reduction in motor automaticity during gait in PD could impair safe mobility as the cortical resources that would be used for compensation are not optimized for the fast, parallel processing required during efficient locomotion.[44]

Furthermore, the excessive attentional demand of walking in PD demands a high computational cost and interferes with gait control during a concurrent cognitive load.[44,46–48] As a result, patients with PD have a greater risk of adverse mobility outcomes and falls, especially during more complex situations where a secondary task is performed in parallel with gait.[46,48,49] In fact, studies have shown larger dual task cost on either gait metrics and/or cognitive measures when a cognitive task is combined with a walking task in people with PD compared with control subjects.[16,44,50]

Dual task paradigms provide further support of PD impairments in automatic control of gait leading to increased variability while walking.[51] When completing gait with a secondary cognitive task, healthy adults typically demonstrate decrements in both the cognitive task (e.g., counting backward by 3s) and gait. These decrements in performance are noted as "dual task cost." Because cognitive tasks are primarily supported by frontal structures and require attention, dual task cost suggests that the primary motor task (e.g., walking) also requires a level of attention and voluntary control. Interestingly, people with PD have more pronounced dual task cost during walking than age-matched adults.[51] This increased dual task cost in people with PD suggests that gait requires more attentional, voluntary control than in healthy adults, resulting in increased variability.[52,53]

In a test of the hypothesis that step time variability is a surrogate for less automatic control of gait, step time variability was measured during a virtual reality walking task while actually peddling in a brain scanner. Consistent with previous studies of overground walking[16,50,51,53,54] patients showed larger step time variability during their dopaminergic Off medication state, than On medication state, consistent with a shift from an automated strategy, toward a more attention-demanding cognitive strategy for controlling gait.[2,43,44] In addition, patients Off dopaminergic medication recruited the cerebellum during periods of increasing gait variability, whereas patients Off medication instead relied upon cortical regions implicated in cognitive control.[55] The cerebellum is a key hub for automated, feed-forward control of motor timing and adaptation.[56–60] Recruiting the cerebellum may have allowed patients on medication to appropriately adapt to increases in step timing variability without the need for excessive attentional control.[57,59,61,62] Without dopaminergic medication, however, the same patients became unable to recruit the cerebellum and instead relied on cortical regions associated with cognitive control (e.g., orbitofrontal cortex).

Fig. 6.6 presents a schematic representation of brain circuitry hypothesized to be involved in automaticity of gait in healthy controls, people with PD On levodopa versus loss of locomotor automaticity in people with PD Off medication (Gilat, 2017). Whereas control subjects, and subjects on levodopa use their basal ganglia together with brainstem and cerebellar locomotor areas to control gait automatically, patients with PD Off levodopa appear to have less automatic control of gait. That is, people with PD Off levodopa tend to use more frontal cortex and attention network connectivities to the basal ganglia to control gait. The slow and serial processing of cortical cognitive areas[63] could have required a longer time for peripheral sensory information to be integrated with the stepping pattern, resulting in a higher step time variability.[44,64,65]

Too low, as well as too high, gait variability may also reflect functional limitations and pathology. Excessive temporal and spatial left-right asymmetry of stepping characteristics has been consistently observed in people with PD.[66] For example, step length[67] and step time[36] have been shown to be more asymmetric in people with PD than in healthy older adults. Temporal and spatial[68] asymmetry of the arms during walking also occurs in people with PD. In fact, one of the earliest signs of abnormal gait in PD is asymmetric arm swing amplitude.[23] These asymmetries of

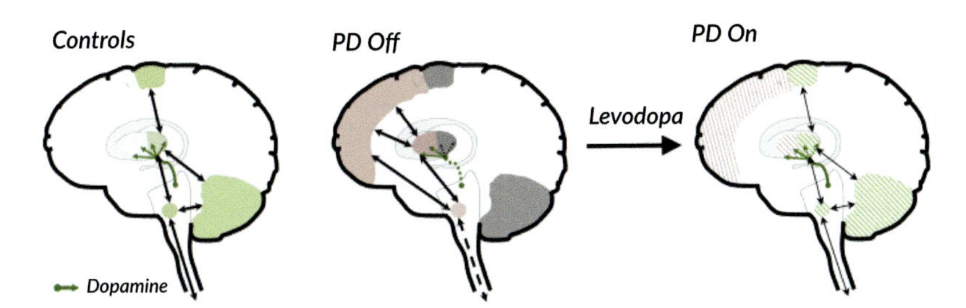

FIGURE 6.6 Schematic representation of the locomotor automaticity processes in healthy adults and people with PD with hypothesized neural mechanisms underlying automaticity. *Left panel*: hypothesized posterior motor network underlying locomotor automaticity in subjects without PD; *Middle panel*: dopaminergic pathology in posterior striatum in PD is thought to cause patients to depend upon attention-demanding, frontal cortical resources to control gait; *Right panel*: dopaminergic replacement therapy is thought to normalize locomotor automaticity impairments in PD. *Adapted from Gilat M, Bell PT, Ehgoetz Martens KA, et al. Dopamine depletion impairs gait automaticity by altering cortico-striatal and cerebellar processing in Parkinson's disease.* NeuroImage *2017;***152***:207–20.*

gait likely relate to the asymmetric onset of bradykinesia and rigidity in the limbs. Furthermore, gait asymmetry, along with variability, has been suggested to lead to more serious gait impairments, such as freezing of gait and falls.[66] Thus, leg and arm asymmetry during gait may represent an important and independent gait impairment in people with PD.

B. How is predictive and reactive obstacle avoidance during walking affected by PD?

Functional mobility requires the ability to predict and avoid or accommodate upcoming obstacles during walking,[69] as well as to react quickly and appropriately when obstacles cannot be avoided.[70] Studies suggest that people with PD take longer to adapt reactive, but not predictive, strategies for changes in the locomotor task based on environmental obstacles or cues.[71]

For example, one study investigated the ability to control the body CoM when adapting to a surface slip while walking as well as adapting to use of a cane.[70] The study showed that cane use improved postural recovery from the first untrained slip, characterized by smaller lateral CoM displacement, in the PD group but not in the control group, who was not as unstabilized by the slip. The beneficial effect of a cane in response to slips occurred only during the first perturbation, and those individuals in the PD group who demonstrated the largest postural displacement without a cane benefited the most from use of a cane. Both PD and control groups gradually decreased lateral CoM displacement across slip exposures, but a slower learning rate was evident in the PD group participants, who required 6, rather than 3, trials for adapting balance recovery (Fig. 6.7).

Another study investigated gait adaptation with a covered exchangeable floor surface to induce gait perturbations.[3] The protocol included a baseline walk on a hard surface, an unexpected trial on a soft surface, and an adaptation phase with 5 soft trials to quantify the reactive adaptation. After the first and sixth soft trials, the surface was changed to hard, to examine after-effects and, thus, predictive motor control. Dynamic stability was assessed using the extrapolated CoM. Patients' unperturbed walking was less stable than controls' and this persisted in the perturbed trials. PD patients did not improve their reactive behavior (e.g., widening their base of support) after repeated perturbations while controls showed clear locomotor adaptation. However, both the PD and control groups demonstrated after-effects immediately after the first perturbation, showing similar predictive responses.[3]

Adaptation of gait termination may also be impaired in people with PD; a study reported results of gait termination adaptation on a slippery surface under unexpected and expected (cued) circumstances. An unexpected slip perturbation during gait termination was under two

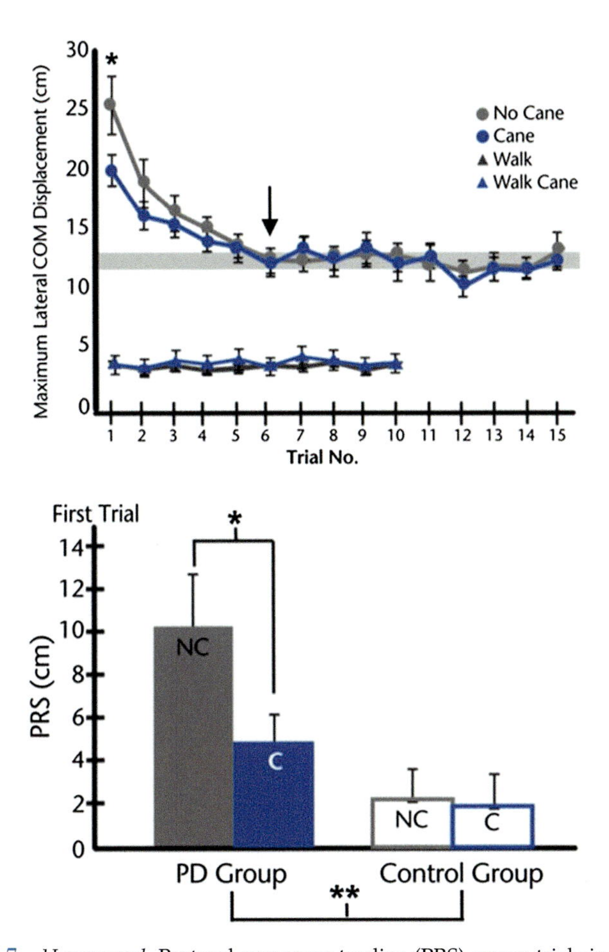

FIGURE 6.7 *Upper panel.* Postural responses to slips (PRS) across trials in participants with PD. Group average of maximum lateral center of mass (CoM) displacement (with standard error of the mean) across 10 trials of unperturbed walking with cane use (walk cane) and without cane use (walk) compared with group average lateral CoM displacement across 15 trials of perturbed gait with and without cane use. The gray zone represents 95% confidence interval of lateral CoM displacement in the 8th to 15th trials, with the arrow indicating the first trial that reached the gray zone. The asterisk indicates a significant difference of the CoM displacement in the first trial from the other trials at $P < .05$. *Lower panel.* Group average PRS (with standard error of the mean) during the first trial of perturbed gait with cane (C) and without cane (NC) in the PD and control groups. The asterisk indicates difference between NC and C trials at $P < .05$ in the PD group. The double asterisk indicates difference in PRS between the PD and control groups at $P < .05$. *Adapted from Boonsinsukh R, Saengsirisuwan V, Carlson-Kuhta P, Horak FB. A cane improves postural recovery from an unpracticed slip during walking in people with Parkinson disease.* Physical Therapy *2012;***92***(9):1117–29.*

conditions: planned over multiple steps and cued one step prior to gait termination. Similar to the control group, the PD group adapted and integrated feed-forward and feed-back components of gait termination under both stop conditions. Feed-forward adaptations included a shorter, wider step, and appropriate stability margin modifications. Feed-back adaptations included a longer, wider subsequent step. When cued to stop quickly, both groups maintained most of these adaptations: foot angle at contact increased in the first cued stop but adapted with practice. However, the group with PD adapted gait terminations with slower, wider steps and less stability.[71]

C. How does levodopa and deep brain stimulation affect dynamic balance, as well as gait?

In general, both levodopa and DBS improve many aspects of straight ahead gait, especially those gait characteristics most related to bradykinesia and rigidity, see Fig. 6.8. In contrast, gait indicators of dynamic balance control are less responsive or even worsen with these therapies, consistent with the observation that falls are not decreased, and may increase when people with PD are in the ON levodopa state and after DBS surgery.[72]

Walking domains	Levodopa	DBS	Rehabilitation
Pace (Stride length, gait speed, etc..)	↑	↑	↑
Variability (Stride time SD, Stance time SD, etc..)	↑	?	?
Rhythm (Stride time, Swing time, Stance time)	▬	?	↑
Asymmetry (Step time asymmetry, etc..)	?	?	?
Balance (Stride width, trunk motion, etc..)	↓	?	↑

FIGURE 6.8 Summary of the effects of Levodopa, DBS, and rehabilitation on different walking domains. Green up arrows indicate improvements, red down arrows indicate worsening, rectangle no change and the question marks indicate not reported in literature.

Levodopa effects

A study of 104 subjects with PD investigated the effects of levodopa on many walking and standing postural sway measures from six body-worn inertial sensors.[13] Fig. 6.9 compares the relative amount of improvement of gait measures (*rightward bars*) and the relative amount of worsening (*leftward bars*). Most improved when people with PD took their levodopa replacement medications were arm swing and leg range of motion, stride velocity and length and peak arm swing velocity. In contrast, signs of

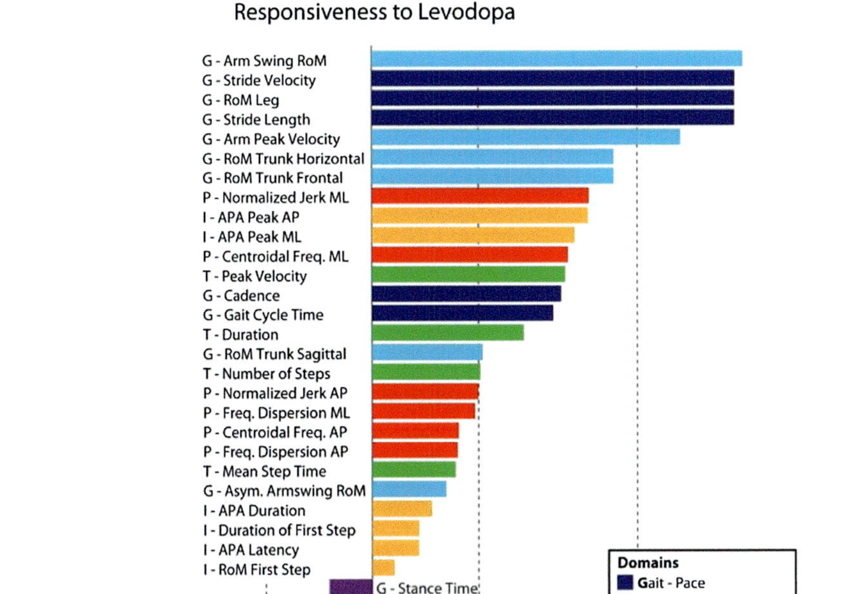

FIGURE 6.9 Responsiveness of balance and gait measures to levodopa medication in 104 subjects with PD. Metrics are ordered from strongest improvement to strongest worsening. PIGD is given as reference in gray. A value larger than 0.20 represents small, 0.50 moderate, and 0.80 large responsiveness. *Adapted from Curtze C, Nutt JG, Carlson-Kuhta P, Mancini M, Horak FB. Levodopa is a double-edged sword for balance and gait in people with Parkinson's disease.* Movement Disorders: Official Journal of the Movement Disorder Society *2015;**30**(10): 1361–70.*

dynamic balance during gait, such as double support time (*yellow*) and turning (*green*) did not significantly change with levodopa. Many measures of postural sway during stance worsened (*Red* in Fig. 6.9, see Chapter 3). In addition to improving bradykinesia of gait, levodopa may improve (reduce) excessive gait variability and asymmetry.[9]

Deep brain stimulation

Like levodopa, DBS in STN or GPi has been shown to improve the bradykinetic (pace) characteristics of parkinsonian gait, with an increase in step length, gait velocity, angular leg excursion, and reduced double-stance duration.[73] STN-DBS also reduces the spatial foot position asymmetry, stride-to-stride variability and interlimb coordination, with a more physiological alternating gait cycle.[74–78] The combination of levodopa treatment and STN-DBS produces an even greater increase in gait velocity than either levodopa or DBS alone.[73] DBS in the GPi also has been reported to significantly increase gait velocity and decrease the double-support duration.[79,80] In contrast, PPN-DBS or SNr-DBS alone resulted in no significant improvements in gait velocity or upper and lower limb movements during gait.[81]

Despite improvements in the speed of walking, there is no evidence that DBS in people with PD improves dynamic postural stability during walking. Temporal measures of gait such as cadence, single and double stance time generally don't change. In addition, pelvis displacement, representing body CoM control, does not improve or can worsen with DBS.[73]

Although DBS of the SNr or PPN, alone, have no effect on gait parameters, they may improve various aspects of balance control.[73] In fact, pace measures of gait, such as stride length and gait velocity are likely controlled by separate neural systems than dynamic balance control of gait, such as braking capacity and trunk displacements. The SNr, a major basal ganglia output known to project to the pedunculopontine nucleus (PPN) in the brainstem is thought to be more involved in balance control whereas the basal ganglia output nuclei, STN and GPi may be more involved in control of the pace of gait although few studies of effects of DBS in the SNr on objective measures of balance in PD are available.[73]

The frequency of DBS may also affect how DBS impacts gait. Whereas high frequency (130 Hz) DBS in the STN can impair dual-task cost or cognitive interference with gait, low frequency stimulation (80 Hz) did not impair dual-task cost.[82] Similar results have been found for freezing of gait. Whereas high frequency stimulation increases freezing of gait (FoG), while improving bradykinesia and rigidity and tremor, lower frequency stimulation does not increase freezing.[83] However, both high and low frequency stimulation have been shown to improve gait speed although neither high nor low frequencies of DBS in STN improve balance control.[78]

D. Rehabilitation and exercise improve gait but do they also improve dynamic balance control during walking?

Over the past 20 years, numerous papers have reported high-quality clinical trials of the effects of physical therapy and group exercises on gait of patients with PD. Meta-analyses of these papers have demonstrated the short-term and, to a lesser extent, long-term benefits of exercise interventions on gait.[84,85] However, the majority of papers focus on gait speed outcomes so the effects of dynamic stability during gait, turning, and more complex tasks are unreported. Nevertheless, the positive outcomes of physical therapy and exercise related to clinical balance scales and fall reduction are consistent with improvement in dynamic balance control during walking as well. In fact, a recent meta-analysis showed that exercise training could decrease the fall rates of PD participants by about 60% over both the short and long terms, although it did not reduce the number of fallers. Metaanalysis results also showed that training at facilities led to more improvement in balance and gait ability over the long term than the community- and home-based training. Facility-based training, mostly supervised by physical therapists, could have enabled participants to practice the training tasks appropriately, at their optimal capacity.

Systematic reviews and clinical guidelines for gait training by physical therapists include enhancing muscle strength; improving aerobic capacity, balance and gait via external cueing; cognitive movement strategies; and physical exercises.[84–86] In a review of the effectiveness of physical therapy (PT) interventions, Tomlinson et al.[85] reported the short-term benefits of physical therapy especially on balance and gait performance, but they could not determine which PT approach is the most effective. Most of the clinical trials and reviews demonstrated only the short-term benefits of PT and exercise interventions.[85]

Several studies of intense aerobic conditioning on treadmills demonstrated improved gait pace performance such as walking speed and stride length as well as gait stability as reflected by less gait variability, and shorter double support time. Both forced (treadmill) and voluntary intense aerobic exercises improve turning speed, gait speed, and stand-sit times, suggesting that intense, but not aerobic, exercise is effective in improving functional aspects of mobility that are often associated with falls and quality of life measures, in addition to gait speed.[87] In contrast to high intensity aerobic training, there is no evidence that endurance exercise training improves gait or Parkinson motor signs, although it does enhance cardiorespiratory capacity and endurance by improving VO2 max.[88]

Auditory and visual cueing during physical therapy gait training and these results were sustained for 2–3 months post-training.[89] However, internal cueing, such as singing, may be more effective than external cueing to improve postural stability during gait in PD.[90] External cues (visual, auditory, and/or vibrotactile) are thought to improve gait by bypassing the basal ganglia that are responsible for internal cueing for gait. Sensory cues likely use cerebellar-cortical and cerebellar-subcortical pathways to trigger step initiation.[91] People with PD who have very short step length and/or freezing of gait have been shown to greatly increase their steps over cues, but only if the cues appear to be moving with respect to themselves, and not if they are stationary due to a strobe light.[92] External cues that improve feedback about walking movements may also be helpful because they can substitute for missing or distorted kinesthetic information about body position or movement. We found that subjects with PD would undershoot stepping onto a visual cue on the floor if they could not see their legs due to a collar around their necks.[93] However, they could accurately increase stepping length to the visual cue when they could see their legs.[93]

Most studies of gait training overground or with a treadmill, with and without body support, demonstrated that the training improved gait speed, stride/step length, and single/double leg stance time ratio, and increased walking capacity as measured by the 6MWT.[94] A minireview[95] reported significant short-term effects of treadmill training on gait hypokinesia and another review found 3–6 months of positive carryover effects in gait performance and mobility.[84] However, no studies have demonstrated improvements in dynamic balance control from gait training. That is, improvements in double support time, trunk range, temporal-spatial variability, turning speed have not been reported with gait training.

Even without gait training, balance training (such as Tango dancing or Tai Chi adapted for PD) has been shown to improve gait speed, motor function, and quality of life of people with PD and benefits are retained 2–6 months after completion of treatment.[86] A subsequent review reported similar positive long-term effects after balance and gait training, as well as a 26%–85% reduction in fall rates at 1–12 month followups.[84] A minimum of 8 weeks of supervised balance training is needed to improve gait performance and reduce fall rates; these effects can be maintained for up to12 months after treatment ends. Balance training has the longest carry-over effects on improving gait, followed by gait training, Tai chi, and cued gait training. Balance training improves balance, gait, and mobility; and it reduces falls for up to 12 months after completion of treatment. Gait training, alone, is effective for improving gait performance and walking capacity up to 6 months post-training, but not for improving balance or falls. Tai chi balance training reduced falls for up to 6 months after treatment ended. Cueing strategies enhance gait speed up to 3 months post-training. Most of the progressive resistive training and aerobic training programs also yield positive effects on gait speed that last for 12 weeks.[86] Unfortunately, few, if any, studies have investigated the effects of rehabilitation on a wide assortment of gait measures such as timing, upper body dynamic balance, asymmetry, and variability on both gait and balance. Most focus on gait speed, alone.

In summary, gait training on treadmills or overground, dynamic balance training such as Tai chi or dance, and/or aerobic exercise on bicycles have been shown to improve gait speed and associated measures of bradykinetic gait. Few studies have investigated whether measures of dynamic balance during walking have been improved with exercise, but many types of gait training when unsupported potentially also train balance control and fall incidence can be reduced, consistent with improvements of both gait speed and balance during gait.

Highlights

- Parkinsonian balance impairments constrain gait and parkinsonian gait impairments constrain balance in many ways.
- Large gait variability in people with PD is thought to reflect less automaticity of control by the nervous system.
- Postural reactions during walking are more affected than predictive obstacle avoidance in people with PD.
- Levodopa and DBS in STN and GPi improve bradykinesia of gait but have little, if any, effect on dynamic balance control.
- Rehabilitation and exercise can improve gait speed, and some specific types of exercise may also improve dynamic balance control during walking to reduce falls.

References

1. Lord S, Galna B, Yarnall AJ, et al. Natural history of falls in an incident cohort of Parkinson's disease: early evolution, risk and protective features. *Journal of Neurology* 2017; **264**(11):2268–76.
2. Peterson DS, Horak FB. Neural control of walking in people with parkinsonism. *Physiology* 2016;**31**(2):95–107.
3. Moreno Catala M, Woitalla D, Arampatzis A. Reactive but not predictive locomotor adaptability is impaired in young Parkinson's disease patients. *Gait and Posture* 2016; **48**:177–82.
4. Curtze C, Nutt JG, Carlson-Kuhta P, Mancini M, Horak FB. Objective gait and balance impairments relate to balance confidence and perceived mobility in people with Parkinson disease. *Physical Therapy* 2016;**96**(11):1734–43.
5. Horak FB, Mancini M, Carlson-Kuhta P, Nutt JG, Salarian A. Balance and gait represent independent domains of mobility in Parkinson disease. *Physical Therapy* 2016;**96**(9): 1364–71.
6. Lord S, Galna B, Rochester L. Moving forward on gait measurement: toward a more refined approach. *Movement Disorders: Official Journal of the Movement Disorder Society* 2013;**28**(11):1534–43.
7. Lord S, Galna B, Verghese J, Coleman S, Burn D, Rochester L. Independent domains of gait in older adults and associated motor and nonmotor attributes: validation of a factor

analysis approach. *The Journals of Gerontology Series A, Biological Sciences and Medical Sciences* 2013;**68**(7):820–7.

8. Morris R, Hickey A, Del Din S, Godfrey A, Lord S, Rochester L. A model of free-living gait: a factor analysis in Parkinson's disease. *Gait and Posture* 2017;**52**:68–71.
9. Hausdorff JM. Gait dynamics in Parkinson's disease: common and distinct behavior among stride length, gait variability, and fractal-like scaling. *Chaos* 2009;**19**(2):026113.
10. Morris ME, Iansek R, Matyas TA, Summers JJ. Ability to modulate walking cadence remains intact in Parkinson's disease. *Journal of Neurology Neurosurgery and Psychiatry* 1994;**57**(12):1532–4.
11. Morris ME, Iansek R, Matyas TA, Summers JJ. The pathogenesis of gait hypokinesia in Parkinson's disease. *Brain: A Journal of Neurology* 1994;**117**(Pt 5):1169–81.
12. Giladi N, Balash J. Paroxysmal locomotion gait disturbances in Parkinson's disease. *Neurologia I Neurochirurgia Polska* 2001;**35**(Suppl. 3):57–63.
13. Curtze C, Nutt JG, Carlson-Kuhta P, Mancini M, Horak FB. Levodopa is a double-edged sword for balance and gait in people with Parkinson's disease. *Movement Disorders: Official Journal of the Movement Disorder Society* 2015;**30**(10):1361–70.
14. Baltadjieva R, Giladi N, Gruendlinger L, Peretz C, Hausdorff JM. Marked alterations in the gait timing and rhythmicity of patients with de novo Parkinson's disease. *European Journal of Neuroscience* 2006;**24**(6):1815–20.
15. Hausdorff JM, Cudkowicz ME, Firtion R, Wei JY, Goldberger AL. Gait variability and basal ganglia disorders: stride-to-stride variations of gait cycle timing in Parkinson's disease and Huntington's disease. *Movement Disorders: Official Journal of the Movement Disorder Society* 1998;**13**(3):428–37.
16. Schaafsma JD, Giladi N, Balash Y, Bartels AL, Gurevich T, Hausdorff JM. Gait dynamics in Parkinson's disease: relationship to Parkinsonian features, falls and response to levodopa. *Journal of the Neurological Sciences* 2003;**212**(1–2):47–53.
17. Mellone S, Mancini M, King LA, Horak FB, Chiari L. The quality of turning in Parkinson's disease: a compensatory strategy to prevent postural instability? *Journal of Neuroengineering and Rehabilitation* 2016;**13**:39.
18. Hong M, Perlmutter JS, Earhart GM. A kinematic and electromyographic analysis of turning in people with Parkinson disease. *Neurorehabilitation and Neural Repair* 2009;**23**(2):166–76.
19. Hulbert S, Ashburn A, Robert L, Verheyden G. A narrative review of turning deficits in people with Parkinson's disease. *Disability and Rehabilitation* 2014:1–8.
20. Huxham F, Baker R, Morris ME, Iansek R. Head and trunk rotation during walking turns in Parkinson's disease. *Movement Disorders: Official Journal of the Movement Disorder Society* 2008;**23**(10):1391–7.
21. Visser JE, Voermans NC, Oude Nijhuis LB, et al. Quantification of trunk rotations during turning and walking in Parkinson's disease. *Clinical Neurophysiology: Official Journal of the International Federation of Clinical Neurophysiology* 2007;**118**(7):1602–6.
22. Mancini M, Horak FB, Zampieri C, Carlson-Kuhta P, Nutt JG, Chiari L. Trunk accelerometry reveals postural instability in untreated Parkinson's disease. *Parkinsonism and Related Disorders* 2011;**17**(7):557–62.
23. Zampieri C, Salarian A, Carlson-Kuhta P, Aminian K, Nutt JG, Horak FB. The instrumented timed up and go test: potential outcome measure for disease modifying therapies in Parkinson's disease. *Journal of Neurology Neurosurgery and Psychiatry* 2010;**81**(2):171–6.
24. Mirelman A, Bernad-Elazari H, Thaler A, et al. Arm swing as a potential new prodromal marker of Parkinson's disease. *Movement Disorders: Official Journal of the Movement Disorder Society* 2016;**31**(10):1527–34.

25. Bloem BR, Hausdorff JM, Visser JE, Giladi N. Falls and freezing of gait in Parkinson's disease: a review of two interconnected, episodic phenomena. *Movement Disorders: Official Journal of the Movement Disorder Society* 2004;**19**(8):871–84.
26. Fasano A, Canning CG, Hausdorff JM, Lord S, Rochester L. Falls in Parkinson's disease: a complex and evolving picture. *Movement Disorders: Official Journal of the Movement Disorder Society* 2017;**32**(11):1524–36.
27. Snijders AH, Takakusaki K, Debu B, et al. Physiology of freezing of gait. *Annals of Neurology* 2016;**80**(5):644–59.
28. Kyrdalen IL, Thingstad P, Sandvik L, Ormstad H. Associations between gait speed and well-known fall risk factors among community-dwelling older adults. *Physiotherapy Research International: The Journal for Researchers and Clinicians in Physical Therapy* 2019;**24**(1):e1743.
29. Uemura K, Yamada M, Nagai K, Tanaka B, Mori S, Ichihashi N. Fear of falling is associated with prolonged anticipatory postural adjustment during gait initiation under dual-task conditions in older adults. *Gait and Posture* 2012;**35**(2):282–6.
30. Berardelli A, Rothwell JC, Thompson PD, Hallett M. Pathophysiology of bradykinesia in Parkinson's disease. *Brain: A Journal of Neurology* 2001;**124**(Pt 11):2131–46.
31. Huang YZ, Chang FY, Liu WC, Chuang YF, Chuang LL, Chang YJ. Fatigue and muscle strength involving walking speed in Parkinson's disease: insights for developing rehabilitation strategy for PD. *Neural Plasticity* 2017;**2017**:1941980.
32. Pedersen SW, Oberg B, Larsson LE, Lindval B. Gait analysis, isokinetic muscle strength measurement in patients with Parkinson's disease. *Scandinavian Journal of Rehabilitation Medicine* 1997;**29**(2):67–74.
33. Rebula JR, Ojeda LV, Adamczyk PG, Kuo AD. The stabilizing properties of foot yaw in human walking. *Journal of Biomechanics* 2017;**53**:1–8.
34. Mazzoni P, Shabbott B, Cortes JC. Motor control abnormalities in Parkinson's disease. *Cold Spring Harbor Perspectives in Medicine* 2012;**2**(6):a009282.
35. Hausdorff JM. Gait variability: methods, modeling and meaning. *Journal of Neuroengineering and Rehabilitation* 2005;**2**:19.
36. Galna B, Lord S, Burn DJ, Rochester L. Progression of gait dysfunction in incident Parkinson's disease: impact of medication and phenotype. *Movement Disorders: Official Journal of the Movement Disorder Society* 2015;**30**(3):359–67.
37. Reimann H, Fettrow TD, Thompson ED, Agada P, McFadyen BJ, Jeka JJ. Complementary mechanisms for upright balance during walking. *PLoS One* 2017;**12**(2):e0172215.
38. Collins SH, Kuo AD. Two independent contributions to step variability during overground human walking. *PLoS One* 2013;**8**(8):e73597.
39. Frenkel-Toledo S, Giladi N, Peretz C, Herman T, Gruendlinger L, Hausdorff JM. Effect of gait speed on gait rhythmicity in Parkinson's disease: variability of stride time and swing time respond differently. *Journal of Neuroengineering and Rehabilitation* 2005;**2**:23.
40. Galna B, Lord S, Rochester L. Is gait variability reliable in older adults and Parkinson's disease? Towards an optimal testing protocol. *Gait and Posture* 2013;**37**(4):580–5.
41. Del Din S, Galna B, Godfrey A, et al. Analysis of free-living gait in older adults with and without Parkinson's disease and with and without a history of falls: identifying generic and disease specific characteristics. *The Journals of Gerontology Series A, Biological Sciences and Medical Sciences* 2019;**74**(4):500–6.
42. van den Heuvel L, Lim AS, Visanji NP, et al. Actigraphy detects greater intra-individual variability during gait in non-manifesting LRRK2 mutation carriers. *Journal of Parkinson's Disease* 2018;**8**(1):131–9.
43. Bohnen NI, Jahn K. Imaging: what can it tell us about parkinsonian gait? *Movement Disorders: Official Journal of the Movement Disorder Society* 2013;**28**(11):1492–500.
44. Clark DJ. Automaticity of walking: functional significance, mechanisms, measurement and rehabilitation strategies. *Frontiers in Human Neuroscience* 2015;**9**:246.

45. Wulf G. *Attention and motor skill learning*. Champaign, IL: Human Kinetics; 2007.
46. Wu T, Hallett M, Chan P. Motor automaticity in Parkinson's disease. *Neurobiology of Disease* 2015;**82**:226–34.
47. Lewis SJ, Barker RA. A pathophysiological model of freezing of gait in Parkinson's disease. *Parkinsonism and Related Disorders* 2009;**15**(5):333–8.
48. Lewis SJ, Shine JM. The next step: a common neural mechanism for freezing of gait. *The Neuroscientist: A Review Journal Bringing Neurobiology, Neurology and Psychiatry* 2016;**22**(1):72–82.
49. Strouwen C, Molenaar EA, Munks L, et al. Dual tasking in Parkinson's disease: should we train hazardous behavior? *Expert Review of Neurotherapeutics* 2015;**15**(9):1031–9.
50. Hausdorff JM, Balash J, Giladi N. Effects of cognitive challenge on gait variability in patients with Parkinson's disease. *Journal of Geriatric Psychiatry and Neurology* 2003;**16**(1):53–8.
51. Kelly VE, Eusterbrock AJ, Shumway-Cook A. A review of dual-task walking deficits in people with Parkinson's disease: motor and cognitive contributions, mechanisms, and clinical implications. *Parkinson's Disease* 2012;**2012**:918719.
52. Peterson DS, Plotnik M, Hausdorff JM, Earhart GM. Evidence for a relationship between bilateral coordination during complex gait tasks and freezing of gait in Parkinson's disease. *Parkinsonism and Related Disorders* 2012;**18**(9):1022–6.
53. Plotnik M, Dagan Y, Gurevich T, Giladi N, Hausdorff JM. Effects of cognitive function on gait and dual tasking abilities in patients with Parkinson's disease suffering from motor response fluctuations. *Experimental Brain Research* 2011;**208**(2):169–79.
54. Bryant MS, Rintala DH, Hou JG, Collins RL, Protas EJ. Gait variability in Parkinson's disease: levodopa and walking direction. *Acta Neurologica Scandinavica* 2016;**134**(1):83–6.
55. Gilat M, Bell PT, Ehgoetz Martens KA, et al. Dopamine depletion impairs gait automaticity by altering cortico-striatal and cerebellar processing in Parkinson's disease. *NeuroImage* 2017;**152**:207–20.
56. Czerny C, Rand T, Gstoettner W, Woelfl G, Imhof H, Trattnig S. MR imaging of the inner ear and cerebellopontine angle: comparison of three-dimensional and two-dimensional sequences. *American Journal of Roentgenology* 1998;**170**(3):791–6.
57. Doya K. Complementary roles of basal ganglia and cerebellum in learning and motor control. *Current Opinion in Neurobiology* 2000;**10**(6):732–9.
58. Lang CE, Bastian AJ. Cerebellar subjects show impaired adaptation of anticipatory EMG during catching. *Journal of Neurophysiology* 1999;**82**(5):2108–19.
59. Morton SM, Bastian AJ. Cerebellar contributions to locomotor adaptations during splitbelt treadmill walking. *Journal of Neuroscience: The Official Journal of the Society for Neuroscience* 2006;**26**(36):9107–16.
60. Wu T, Hallett M. The cerebellum in Parkinson's disease. *Brain: A Journal of Neurology* 2013;**136**(Pt 3):696–709.
61. Horak FB, Diener HC. Cerebellar control of postural scaling and central set in stance. *Journal of Neurophysiology* 1994;**72**(2):479–93.
62. Morton SM, Bastian AJ. Cerebellar control of balance and locomotion. *The Neuroscientist: A Review Journal Bringing Neurobiology, Neurology and Psychiatry* 2004;**10**(3):247–59.
63. Schneider W, Chein JM. Controlled & automatic processing: behavior, theory, and biological mechanisms. *Cognitive Science* 2003;**27**(3):525–59.
64. Hamacher D, Herold F, Wiegel P, Hamacher D, Schega L. Brain activity during walking: a systematic review. *Neuroscience and Biobehavioral Reviews* 2015;**57**:310–27.
65. Lucas M, Chaves F, Teixeira S, et al. Time perception impairs sensory-motor integration in Parkinson's disease. *International Archives of Medicine* 2013;**6**(1):39.
66. Plotnik M, Giladi N, Hausdorff JM. A new measure for quantifying the bilateral coordination of human gait: effects of aging and Parkinson's disease. *Experimental Brain Research* 2007;**181**(4):561–70.

67. Roemmich RT, Nocera JR, Stegemoller EL, Hassan A, Okun MS, Hass CJ. Locomotor adaptation and locomotor adaptive learning in Parkinson's disease and normal aging. *Clinical Neurophysiology: Official Journal of the International Federation of Clinical Neurophysiology* 2014;**125**(2):313–9.
68. Lewek MD, Poole R, Johnson J, Halawa O, Huang X. Arm swing magnitude and asymmetry during gait in the early stages of Parkinson's disease. *Gait and Posture* 2010;**31**(2): 256–60.
69. Patla AE, Prentice SD, Robinson C, Neufeld J. Visual control of locomotion: strategies for changing direction and for going over obstacles. *Journal of Experimental Psychology Human Perception and Performance* 1991;**17**(3):603–34.
70. Boonsinsukh R, Saengsirisuwan V, Carlson-Kuhta P, Horak FB. A cane improves postural recovery from an unpracticed slip during walking in people with Parkinson disease. *Physical Therapy* 2012;**92**(9):1117–29.
71. Oates AR, Van Ooteghem K, Frank JS, Patla AE, Horak FB. Adaptation of gait termination on a slippery surface in Parkinson's disease. *Gait and Posture* 2013;**37**(4):516–20.
72. Lilleeng B, Gjerstad M, Baardsen R, Dalen I, Larsen JP. Motor symptoms after deep brain stimulation of the subthalamic nucleus. *Acta Neurologica Scandinavica* 2015;**131**(5): 298–304.
73. Collomb-Clerc A, Welter ML. Effects of deep brain stimulation on balance and gait in patients with Parkinson's disease: a systematic neurophysiological review. *Neurophysiologie Clinique = Clinical Neurophysiology* 2015;**45**(4–5):371–88.
74. Cantiniaux S, Vaugoyeau M, Robert D, et al. Comparative analysis of gait and speech in Parkinson's disease: hypokinetic or dysrhythmic disorders? *Journal of Neurology Neurosurgery and Psychiatry* 2010;**81**(2):177–84.
75. Johnsen EL, Mogensen PH, Sunde NA, Ostergaard K. Improved asymmetry of gait in Parkinson's disease with DBS: gait and postural instability in Parkinson's disease treated with bilateral deep brain stimulation in the subthalamic nucleus. *Movement Disorders: Official Journal of the Movement Disorder Society* 2009;**24**(4):590–7.
76. McNeely ME, Earhart GM. Medication and subthalamic nucleus deep brain stimulation similarly improve balance and complex gait in Parkinson disease. *Parkinsonism and Related Disorders* 2013;**19**(1):86–91.
77. McNeely ME, Hershey T, Campbell MC, et al. Effects of deep brain stimulation of dorsal versus ventral subthalamic nucleus regions on gait and balance in Parkinson's disease. *Journal of Neurology Neurosurgery and Psychiatry* 2011;**82**(11):1250–5.
78. Vallabhajosula S, Haq IU, Hwynn N, et al. Low-frequency versus high-frequency subthalamic nucleus deep brain stimulation on postural control and gait in Parkinson's disease: a quantitative study. *Brain Stimulation* 2015;**8**(1):64–75.
79. Allert N, Volkmann J, Dotse S, Hefter H, Sturm V, Freund HJ. Effects of bilateral pallidal or subthalamic stimulation on gait in advanced Parkinson's disease. *Movement Disorders: Official Journal of the Movement Disorder Society* 2001;**16**(6):1076–85.
80. Defebvre LJ, Krystkowiak P, Blatt JL, et al. Influence of pallidal stimulation and levodopa on gait and preparatory postural adjustments in Parkinson's disease. *Movement Disorders: Official Journal of the Movement Disorder Society* 2002;**17**(1):76–83.
81. Peppe A, Pierantozzi M, Chiavalon C, et al. Deep brain stimulation of the pedunculopontine tegmentum and subthalamic nucleus: effects on gait in Parkinson's disease. *Gait and Posture* 2010;**32**(4):512–8.
82. Varriale P, Collomb-Clerc A, Van Hamme A, et al. Decreasing subthalamic deep brain stimulation frequency reverses cognitive interference during gait initiation in Parkinson's disease. *Clinical Neurophysiology: Official Journal of the International Federation of Clinical Neurophysiology* 2018;**129**(11):2482–91.

83. Moreau C, Defebvre L, Devos D, et al. STN versus PPN-DBS for alleviating freezing of gait: toward a frequency modulation approach? *Movement Disorders: Official Journal of the Movement Disorder Society* 2009;**24**(14):2164–6.
84. Mak MK, Wong-Yu IS, Shen X, Chung CL. Long-term effects of exercise and physical therapy in people with Parkinson disease. *Nature Reviews Neurology* 2017;**13**(11): 689–703.
85. Tomlinson CL, Herd CP, Clarke CE, et al. Physiotherapy for Parkinson's disease: a comparison of techniques. *Cochrane Database of Systematic Reviews* 2014;(6):Cd002815.
86. Shen X, Wong-Yu IS, Mak MK. Effects of exercise on falls, balance, and gait ability in Parkinson's disease: a meta-analysis. *Neurorehabilitation and Neural Repair* 2016;**30**(6): 512–27.
87. Miller Koop M, Rosenfeldt AB, Alberts JL. Mobility improves after high intensity aerobic exercise in individuals with Parkinson's disease. *Journal of the Neurological Sciences* 2019; **399**:187–93.
88. Lamotte G, Rafferty MR, Prodoehl J, et al. Effects of endurance exercise training on the motor and non-motor features of Parkinson's disease: a review. *Journal of Parkinson's Disease* 2015;**5**(1):21–41.
89. Ginis P, Nackaerts E, Nieuwboer A, Heremans E. Cueing for people with Parkinson's disease with freezing of gait: a narrative review of the state-of-the-art and novel perspectives. *Annals of Physical and Rehabilitation Medicine* 2018;**61**(6):407–13.
90. Harrison EC, Horin AP, Earhart GM. Internal cueing improves gait more than external cueing in healthy adults and people with Parkinson disease. *Scientific Reports* 2018;**8**(1): 15525.
91. Mackinnon C. Sensorimotor anatomy of gait, balance, and falls. In: *Handbook of clinical neurology*. Elsevier; 2018.
92. Azulay JP, Mesure S, Amblard B, Blin O, Sangla I, Pouget J. Visual control of locomotion in Parkinson's disease. *Brain: A Journal of Neurology* 1999;**122**(Pt 1):111–20.
93. Jacobs JV, Horak FB. Abnormal proprioceptive-motor integration contributes to hypometric postural responses of subjects with Parkinson's disease. *Neuroscience* 2006; **141**(2):999–1009.
94. Mehrholz J, Kugler J, Storch A, Pohl M, Hirsch K, Elsner B. Treadmill training for patients with Parkinson's disease. *Cochrane Database of Systematic Reviews* 2015;**9**.
95. Herman T, Giladi N, Hausdorff JM. Treadmill training for the treatment of gait disturbances in people with Parkinson's disease: a mini-review. *Journal of Neural Transmission* 2009;**116**(3):307–18.

CHAPTER

7

How and why is turning affected by Parkinson disease?

Clinical case

When Marie was diagnosed with Parkinson disease (PD), she felt her walking was good; however, she noticed that she experienced several near falls when turning in the kitchen while cooking. Levodopa replacement helped increase her gait speed, but she started noticing more stumbles while turning when On medication and more episodes of freezing of gait when Off medication, especially while turning. Her doctor referred her to physical therapy where she practiced turning on a rotating treadmill; however, she felt that more improvements were obtained when combining physical therapy with external cues so she started carrying a small metronome to help with stepping during turning.

A. Why is turning a difficult dynamic balance task?

Turning constitutes an essential part of functional mobility in daily life considering that the majority of activities in the home require three to four turns and over 50% of daily steps are turning steps.[1,2] Thus, people turn nearly 1000 per day, although it is a difficult balance task.[3] Turning while walking is challenging because it consists of decelerating the forward motion, rotating the body, and stepping out toward a new direction. Normally, gaze rotation occurs approximately 200 ms before the start of the turn.[4,5] This so-called "go where you look" strategy[6] occurs also in the dark.[7] After the head, the trunk, later the pelvis and feet rotate to the inner side of the turning cycle, and the center of mass (CoM) deviates to the same side.[8,9] During turning, muscle activation changes, velocity

https://doi.org/10.1016/B978-0-12-813874-8.00007-6

decreases, and stride width increases to improve stability of body weight during trunk rotation and lateral translation, compared to straight gait.[4,10]

Difficulties turning while walking are especially common among people with PD,[11,12] and negatively affect functional independence. Together with gait difficulties, turning difficulties have recently been shown to be a major risk factor for falls, institutionalization, and death in PD.[13] Turning requires the central nervous system to coordinate body reorientation toward a new travel direction while continuing with the ongoing step cycle and maintaining mediolateral stability.[14,15] Turning is more likely vulnerable to functional impairments compared to straight ahead gait, since turning involves more interlimb coordination, more multisensory integration, more coupling between posture and gait, and modification of locomotor patterns requiring frontal lobe cognitive and executive function that play a role in postural transitions.[16,17]

B. PD affects many sensorimotor control systems important for controlling a turn

The reduced step length, increased step number, and alteration of turn strategy might predispose those with PD to experience difficulty in turning. However, it is not clear if these effects are direct responses to the physiological changes as a result of PD pathology, or secondary effects from a need to gain more stability and to compensate for impairments affecting turning. Below, we summarize the different sensorimotor systems impaired in PD and their potential effects on turning (Fig. 7.1).

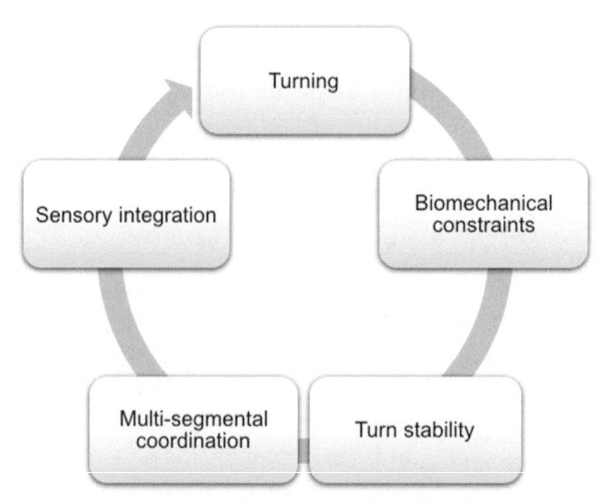

FIGURE 7.1 Different sensorimotor systems involved in turning.

Biomechanical constraints

Turning involves internal/external rotation of the hip joints with many deep muscles involved both with a foot planted onto the floor and during the swing phase of gait.[4,10] Structurally, the axial musculature links all parts of the body, giving support and stability, to allow mobility and proximal limb movement. The axial musculature is anatomically and physiologically complex, controlled by cortical and subcortical structures via the descending monoaminergic pathways (i.e., reticulospinal and vestibulospinal motor pathways). Although a degree of tonic regulation (sustained postural control) of axial muscles is necessary to provide a stable base of support for movement of body segments to be coordinated and controlled, excessive tonic activation as a result of axial motor impairments in people with PD leads to an axial segment rigidity, impairing turning.[18–20]

The first study to assess the ability of people with PD to turn in bed used a Kings College Hospital (KCH) rating scale, with the subscore of "axial rotation" consisting of a mean of the scores rating rising from a chair, postural stability, axial rigidity, and whole body bradykinesia.[20] In total, 19/36 participants demonstrated difficulty in turning in bed when examined in the Off state following a levodopa drug withdrawal. Axial rotation and difficulty turning in bed were correlated, suggesting an influence of axial control on turning performance.

Later, a more quantitative measure of axial tone in standing demonstrated that people with PD have significantly larger axial torque compared to healthy controls[18] (43% more, with the largest increases in hip torque, compared to trunk or neck, as previously described in Chapter 3). In addition, higher neck torque was significantly related to a longer duration of turning in a functional gait task[18] (Fig. 7.2). These results

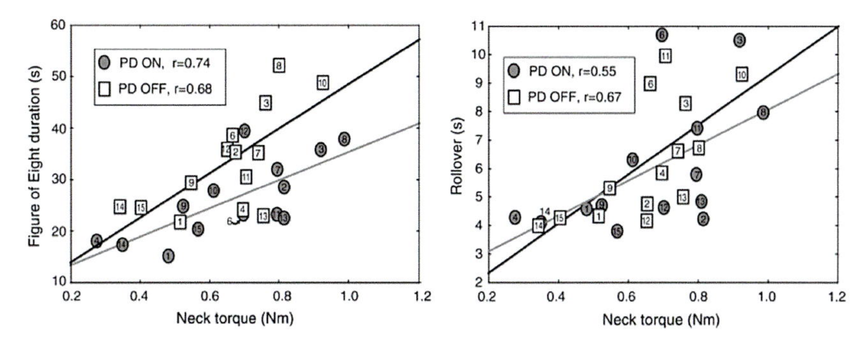

FIGURE 7.2 Relation between neck torque and time needed to perform a Figure of Eight task (*left*) and the rollover (*right*) in both PD OFF (*boxes*) and ON (*circles*). Regression lines are shown for the PD subjects OFF (*solid black*) and ON (*solid gray*) medication. *Adapted from Franzen E, Paquette C, Gurfinkel VS, Cordo PJ, Nutt JG, Horak FB. Reduced performance in balance, walking and turning tasks is associated with increased neck tone in Parkinson's disease.* Experimental Neurology *2009;**219**(2):430–438.*

suggest axial rigidity, leading to deficits of axial rotation, to be a key contributor to the development of turning difficulties experienced by people with PD.

Turn stability

Dynamic stability, that is, the ability to control the body's center of mass (CoM) over its moving base of support, is compromised during a turn since the CoM may momentarily move outside the base of support naturally, increasing the risk for falls,[21–23] (Fig. 7.3).

Generally, subjects with PD show a narrower base of support than control subjects while turning at their self-selected speeds. In fact, the minimum distance between the ankles while turning is always smaller in subjects with PD compared to healthy control subjects.[24] The narrower base of support during turns might result in worse dynamic stability, that is, the mean distance between the body CoM and the lateral margin of the foot is often smaller in the PD, compared to the control group.[21] In fact, with a narrower base of support it would be easier for the CoM to fall outside the feet while turning.

Dynamic stability, measured as the mean distance between the CoM (or extrapolated CoM (ECOM), which includes CoM velocity) and the lateral margin of the feet was significantly smaller in subjects with PD.[25]

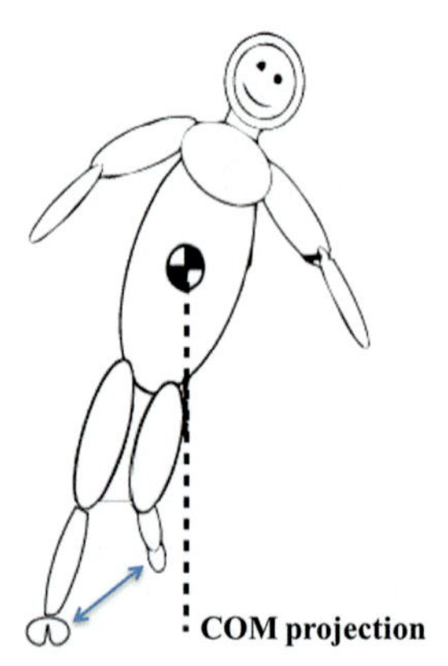

FIGURE 7.3 Schematic of the center of mass (CoM) falling outside the base of support while turning. *Adapted from Fino PC, Lockhart TE, Fino NF. Corner height influences center of mass kinematics and path trajectory during turning.* Journal of Biomechanics *2015;**48**(1): 104–112.*

Interestingly, subjects with PD significantly spend more time with the CoM (or ECOM) outside the lateral base of support during turning 90 degrees at fast matched speeds (>65 cm/s) compared to control subjects[25] (Fig. 7.4).

It is interesting to notice that the most unstable turns occurred at 90 degrees, which is the most common turning angle made during daily activities.[26] Thus, fall prevention rehabilitation could focus on modifying turning strategies, particularly for 90 degrees turns, such as practicing a wider arc turn, turning more slowly or practicing sustaining a wider base of support. Scaling turning speeds for various speeds of walking is another important characteristic to train in order to reduce falls. People with PD may lose their balance due to inability to change turning strategy with change in motor commands for a change in speed. On the other hand, subjects with PD might be actively slowing and widening their turns to compensate for postural instability due to narrower base of support but more evidence is needed to support this hypothesis.[25]

Turn strategy

People with PD have been shown to predominantly use an incremental turn type (turning on-the-spot before walking) as opposed to a

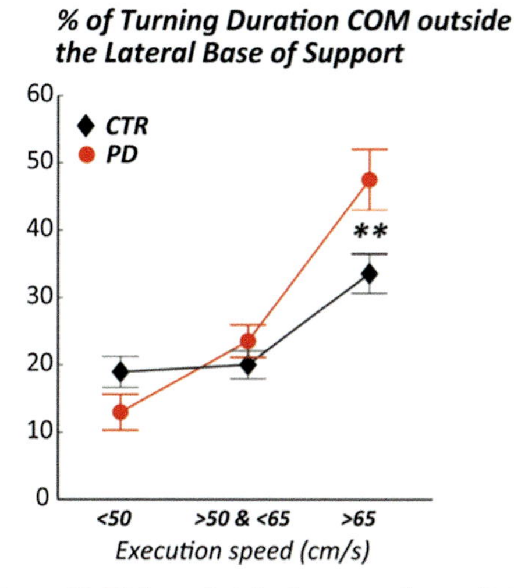

FIGURE 7.4 Subjects with PD have their body center of mass (CoM) (extrapolated CoM) outside their base of support for longer duration during fast turns. Group means (±SEM) for the % of turning duration in which CoM falls outside the lateral base of support in 90 degrees turns and matched speeds. ** $P < .01$. *Adapted from Mellone S, Mancini M, King LA, Horak FB, Chiari L. The quality of turning in Parkinson's disease: a compensatory strategy to prevent postural instability?* Journal of Neuroengineering and Rehabilitation *2016;***13***:39.*

toward-type turn (direct advance to target while turning) or a pivot-type turn (broad-base turn initiating advance to target) used by healthy controls.[12,27,28] If the turn is not to the favored direction, however, both controls and people with PD perform a turn in the unfavored direction with an "incremental turn" strategy in which a turn on the spot is followed by walking.[12] People with greater motor symptoms (as measured by their UPDRS Score) and reduced balance confidence (Activities-specific Balance Confidence scale) adopt a multiple stepping turn pattern rather than a spin turn, supporting the notion that potential instability causes an adaptive change in turn strategy.[12]

People with PD tend to use a step-round style (by stepping around without a pivot foot) rather than a spin style (by pivoting around a foot) used by healthy controls.[29] This study also reported the step-round strategy to result in greater postural stability (so less amount of time with the CoM outside the base of support), consistent with a safer, adaptive turning strategy.[29]

These results suggest dysfunctional turning in people with PD may be a result of a lack of perceived stability requiring a slowing and adaption of their turning style. This change in turn strategy may be compensatory to gain control of the CoM within the base of support, secondary to impaired postural control and inability to adjust to the functional task demands.

Multisegmental coordination

Literature on turning in PD highlights the importance of altered oculomotor control, noting that individuals with PD make fewer preparatory saccades approaching a turn[30] and initiate turns with saccades that are slower, smaller, and more frequent than age-matched control subjects,[31] see Fig. 7.5.

People with PD also demonstrate slowness of head and trunk reorientation movements which may be compensated by greater contribution of eye movements than of head/trunk movements to achieve gaze shifts associated with turning.[32] In fact, the characteristics of the saccade initiating a turn are predictive of the turn performance; turns initiated with larger, faster saccades are executed more quickly than turns initiated with smaller, slower saccades.[31] Since people with PD have smaller, slower saccades, their turns are also smaller and slower.

Healthy young and older adults use a craniocaudal sequence of body rotations to turn while walking, with head rotation leading, then trunk and then pelvis rotation to reorientate the body toward the new direction[4,10,33] resulting in intersegmental reciprocal movements.[14] In contrast, people with PD demonstrate a loss of axial rotation of the spine[34] with little dissociation between the head, trunk, and lower limbs during

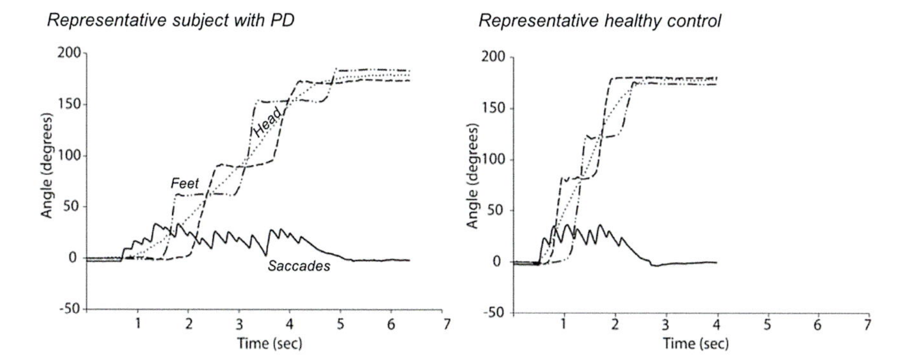

FIGURE 7.5 Representative data from individual turn trials showing eye, head, and foot rotations in the horizontal plane. *Left panel*: Representative 180-degree turn performed by an individual with PD. The subject performed 15 saccades of varying amplitudes during the turn, and required five steps to complete the turn. *Right panel*: Representative 180-degree turn performed by a healthy control. The subject performed 8 saccades of more consistent amplitude than those performed by the individual with PD, and required only four steps and less time to complete the turn than the individual with PD. *Adapted from Lohnes CA, Earhart GM. Saccadic eye movements are related to turning performance in Parkinson disease.* Journal of Parkinson's Disease *2011;**1**(1):109–118.*

turning. This lack of intersegmental rotation coordination results in an "en-bloc" turn,[8,33] that is, a turn with a nearly simultaneous rotation of head and trunk. Fig. 7.6 shows that the onset of head rotation leads the onset of upper trunk rotation by approximately 300 ms in a control subject (*left*) but no delay between head and trunk rotation in a subject with PD.

Kinematic analysis of body segment rotations during turning reveal clear abnormalities in people with PD, including head-on-trunk turning,[18,33] trunk-on-hip turning,[35,36] and whole body turning.[15,19]

Not only is the coordination of the timing of body segments during the onset of turn affected by PD, but also the velocity of the segmental rotation during the turn is slow so the total time to complete the turn is longer than normal.[1] Slow turning rotation may be a result of the bradykinesia experienced by people with PD.[37] However, PD with relatively unaffected walking speed still present with slow, "en-bloc" turns, often similar to those with more advanced symptoms. In fact, walking velocity is unrelated to segmental coordination, suggesting that bradykinetic gait is unlikely to be the sole driver of axial coordination deficits in people with PD. It has been suggested[8] that adopting an "en-bloc" turning strategy with less degrees of freedom simplifies the turn and therefore is adopted by people with PD in response to a lack of neural control over multisegmental coordination. In fact, healthy, age-matched control subjects also tend to turn "en-bloc" under challenging conditions.[38]

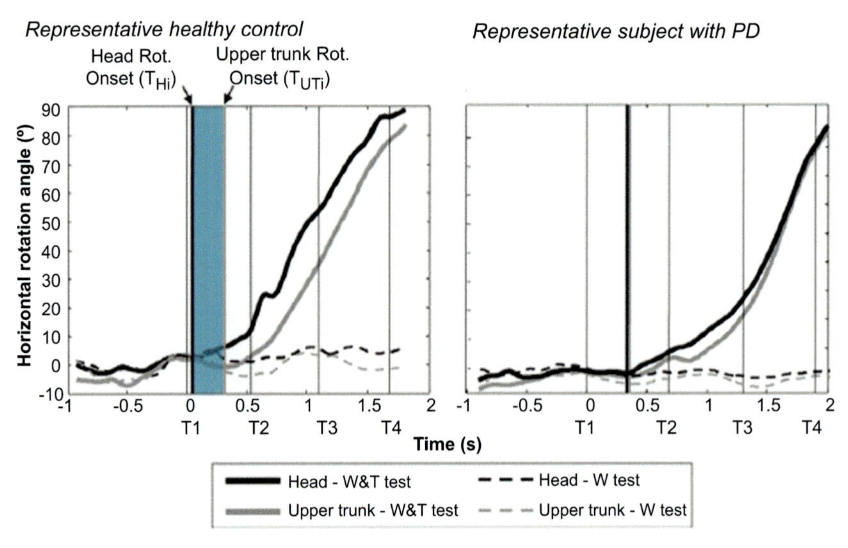

FIGURE 7.6 Time courses of head (*black*) and upper trunk (*gray*) horizontal rotations during a straight walking (*dotted lines*) and a turning (*bold lines*) test for a control subject (*left*) and a representative patient with early PD (*right*). *Adapted from Crenna P, Carpinella I, Rabuffetti M, et al. The association between impaired turning and normal straight walking in Parkinson's disease.* Gait and Posture *2007;**26**(2):172–178.*

Sensory integration

To effectively execute a turn, one must not only generate a motor pattern appropriate for turning but also integrate sensory information from several sources to determine how far one has turned. The role of sensory information in perception of turning has been the focus of several studies in healthy controls.[39–43] These studies have shown that people are less accurate at estimating displacements when passively turned than when actively turning themselves.[40] During passive turning, primarily vestibular cues are available, and cognitive processes compensate for the decay of vestibular cues.[41] During active turning, vestibular, proprioceptive, and efference copy cues are available. Efference copy cues are duplicates of the outgoing motor command that are relayed centrally to predict expected sensory feedback. This fusion of vestibular, proprioceptive, and efference copy cues may improve accuracy of active self-turning compared to passive turning. Deficits in fusion of information from different systems and in individual sensory system function could lead to misperception of turning. However, that is not the case in PD. In fact, a study from Earhart[44] found subjects with PD are just as accurate as controls in estimating turn amplitudes in both active and passive conditions with or without vision in people with mild PD tested on their medications. Furthermore, people with mild PD show no asymmetry in ability to estimate turn amplitudes despite asymmetry in the Parkinsonian symptoms. Therefore, difficulties with turning in PD are likely to be related to motor impairments or deficits in sensorimotor integration, not primarily perceptual impairments.

C. Turning impairments are sensitive to early disease, falls, and freezing of gait

In summary, turning problems are an early sign in mild PD, as recently summarized in a narrative review by Hulbert et al.[38] Compared to age-matched controls, people with PD need more time and steps to complete a turn. Turning is characterized by a broader turning arc and step length is even more reduced during tighter turns. People with PD are less likely to take crossover steps especially with increasing turning angles. Besides spatiotemporal impairments, the initiation of head orientation is postponed in PD, resulting in a more coupled start of head and trunk rotation in comparison with the top-down coordination in healthy controls. Furthermore, turning is characterized by an increased forward inclination and instability of the CoM trajectory in people with PD.

Turning deficits such as a large number of steps and slow velocity are present even in early stages of PD, when patients are not yet taking antiparkinson medication and when their walking speed is similar to healthy controls.[45] In fact, Zampieri et al.[45] found early, untreated people with PD to take similar total time to complete a 3 or 7 m, instrumented Timed-Up and Go test but with significantly slow turning, see Fig. 7.7. Similarly, another study highlighted the importance of quantitative measures of turning to discriminate people with mild PD from elderly controls, as opposed to common clinical scales of balance such as the Tinetti or the Berg Balance Scale.[17]

Turning impairments have been found to be related to perceived quality of life, as measured by the PDQ-39[46] and to disease severity, as measured by the UPDRS.[11] Although an increase in step number is believed to indicate difficulty in turning, being less efficient and of lower quality,[11] it is possible that a purposeful increase in steps during a turn may, in fact, be useful to maintain functional stability. People with PD who have a history of falls or have severe PD use this compensatory strategy of many small steps to turn.

Videos of people with PD during daily life suggest that 70% of falls in individuals with PD were intrinsic in nature and often occurred during transitional activities such as turning around.[47] Studies from Weaver[48] and Robinovitch[49] examined the causes and circumstances of falls in

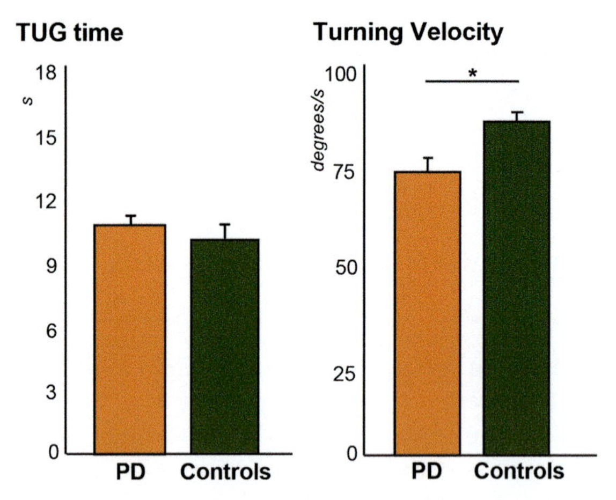

FIGURE 7.7 Comparison in TUG time between untreated subjects with PD and control subjects on the traditional, 3 m TUG test, and Turning Velocity measured with inertial sensors during an instrumented, 7 m TUG test. Vertical bars are standard errors. *Adapted from Zampieri C, Salarian A, Carlson-Kuhta P, Aminian K, Nutt JG, Horak FB. The instrumented timed up and go test: potential outcome measure for disease modifying therapies in Parkinson's disease.* Journal of Neurology, Neurosurgery, and Psychiatry *2010;**81**(2):171–176.*

individuals with PD, as well as in older adults, through the analysis of real-life falls captured on video. In the older group[49] incorrect weight shifting accounted for 46.4% of falls (410 falls), followed by loss of support from an external object (e.g., cane, walker) (20.8%, 185 falls), and trip or stumble (13.3%, 117 falls). Walking was the most common activity at the time of falling, accounting for 36.3% of cases (321 falls), followed by standing still (24.2%, 214 falls), and transferring from standing (17.3%, 153 falls). People with PD were 1.3 times as likely as those without PD to fall because of incorrect weight shifting[48]; 61% of falls in individuals with PD were due to incorrect weight shifting, as would occur during turning and other postural transitions, compared with 46% in those without PD.

Overall, larger and sharper turns are associated with increased step time variability and more freezing in individuals with PD and a history of freezing of gait (FoG).[27,50,51] In addition, people with PD and FoG tend to turn with a longer turn duration, slower peak turn velocity, more steps, and faster cadence compared to nonfreezers when recorded during daily life activities.[52] Sharp, large turning angles are considered an important trigger for this abnormal, slow pattern, so people with PD who have a tendency to freeze tend to avoid sharp, large turns.

D. Can levodopa, deep brain stimulation, or exercise improve turning?

Although dopaminergic medication has a dramatic clinical effect on motor impairments in PD,[53] the effects of medication on turning remain uncertain and complex.[54,55] For instance, while positive effects of medication on global measures of straight-ahead gait in individuals with PD is well documented (see Chapter 6) negative effects on postural sway (see Chapter 3) and on automatic postural responses (see Chapter 4) have also been reported.

Few studies have examined the role of dopaminergic medication in turning.[55–57] In general, after intake of dopaminergic medication, subjects with PD increased turning speed, decreased turning duration and turning distance but not the amount of body rotation. Despite the significant effect of dopaminergic medication, turning characteristics did not improve in individuals with PD to the level of the control group.[55–57] Specifically, the persisting impairments are related to poor effects of medication on step width and step length.

Problems regulating step width, not step length, have been documented in PD.[38] Specifically, individuals with PD turn with narrower steps, sometimes using a crossing step (i.e., step width closer to a value of zero).[24,56,58] Crossing steps while turning is not only a complex motor task that could induce instability due to the drastic change of the base of support, but is also an important contributor to body rotation.[58] Consequently, problems modulating step width in PD compromise mediolateral stability[59,60] and compromise force production necessary to accelerate the CoM toward the turn direction.[24] These findings highlight the importance of targeting mediolateral stability during rehabilitation sessions in individuals with PD. In addition, these findings can partly be explained by the ineffectiveness of levodopa in improving axial deficits and axial tone.[18,19]

One small study on eleven patients with PD reported the effects of STN DBS on turning.[61] DBS increased the amplitude, velocity, and main-sequence-slope of the first saccade performed during the turn.[61] In addition, the intersegmental latencies between the eyes, head, trunk, and feet were increased by DBS; however, this positive effect was eliminated when controlled for turning speed.[61] These findings on improved turning duration and concomitant improvements in oculomotor performance during turns are in keeping with other studies showing efficacy of DBS in improving saccade function and gait. In contrast, other studies show DBS is detrimental for many aspects of balance control.[62] The choice of patients for DBS surgery may also play a role here. In fact, the subjects in the study showing improved multisegmental delays during turning did not show an abnormal, en-bloc turning strategy prior to DBS.[61] Therefore, more studies exploring turning strategies before and after DBS are needed before considering DBS for turning impairments.

Little evidence points to an improvement in turning performance in people with PD with rehabilitation intervention. Results should be cautiously interpreted due to small sample size in the studies. Specifically, 12 sessions of curved-walking training (by a rotating treadmill) provided the first evidence of turning improvements after rehabilitation in 12 subjects with PD.[63] However, only step length, speed, and cadence during turning were examined, with no mention of other turning characteristics that tend to be nonresponsive to dopaminergic medication.[63] In addition, it isn't clear if a regular linear treadmill would provide the same type of improvement as a rotating treadmill. In fact, another study examined the effects of linear treadmill rehabilitation as well as an agility boot camp training[64] and showed similar improvements of both interventions on shortening turning duration. Lastly, rhythmical cueing yielded faster performance of a functional turn in people with PD with or without a history of freezing of gait in a cohort of 133 patients.[65,66] Whether this constitutes a robust clinical improvement needs further confirmation by analysis of possible harmful effects of cueing, such as increasing the risk of freezing and falls.

Highlights

- Many sensorimotor systems necessary for turning are impaired in PD.
- Higher neck tone is related to worse turning in PD.
- Fast turning may be more unstable than slow turning in PD.
- An "en-bloc" turning strategy is common in PD opposed to a normal, top-down multisegmental turning strategy.
- Turning is impaired even when walking speed is normal in PD.
- Weight-shifting, as necessary during turning, has been found to be the most common fall situation.
- Levodopa seems to improve only certain aspects of turning and never restores values to control levels.
- DBS effects on turning are not yet conclusive.
- Rehabilitation and cueing may be beneficial for turning; the most effective rehabilitation is not known.

References

1. Earhart GM. Dynamic control of posture across locomotor tasks. *Movement Disorders: Official Journal of the Movement Disorder Society* 2013;**28**(11):1501–8.
2. Glaister BC, Bernatz GC, Klute GK, Orendurff MS. Video task analysis of turning during activities of daily living. *Gait and Posture* 2007;**25**(2):289–94.

3. Mancini M, Schlueter H, El-Gohary M, et al. Continuous monitoring of turning mobility and its association to falls and cognitive function: a pilot study. *The Journals of Gerontology Series A, Biological Sciences and Medical Sciences* 2016;**71**(8):1102–8.
4. Courtine G, Schieppati M. Human walking along a curved path. II. Gait features and EMG patterns. *European Journal of Neuroscience* 2003;**18**(1):191–205.
5. Hicheur H, Vieilledent S, Berthoz A. Head motion in humans alternating between straight and curved walking path: combination of stabilizing and anticipatory orienting mechanisms. *Neuroscience Letters* 2005;**383**(1–2):87–92.
6. Grasso R, Prevost P, Ivanenko YP, Berthoz A. Eye-head coordination for the steering of locomotion in humans: an anticipatory synergy. *Neuroscience Letters* 1998;**253**(2):115–8.
7. Authie CN, Hilt PM, N'Guyen S, Berthoz A, Bennequin D. Differences in gaze anticipation for locomotion with and without vision. *Frontiers in Human Neuroscience* 2015;**9**:312.
8. Crenna P, Carpinella I, Rabuffetti M, et al. The association between impaired turning and normal straight walking in Parkinson's disease. *Gait and Posture* 2007;**26**(2):172–8.
9. Pradeep Ambati VN, Murray NG, Saucedo F, Powell DW, Reed-Jones RJ. Constraining eye movement when redirecting walking trajectories alters turning control in healthy young adults. *Experimental Brain Research* 2013;**226**(4):549–56.
10. Patla AE, Adkin A, Ballard T. Online steering: coordination and control of body center of mass, head and body reorientation. *Experimental Brain Research* 1999;**129**(4):629–34.
11. Stack E, Ashburn A. Dysfunctional turning in Parkinson's disease. *Disability and Rehabilitation* 2008;**30**(16):1222–9.
12. Stack EL, Ashburn AM, Jupp KE. Strategies used by people with Parkinson's disease who report difficulty turning. *Parkinsonism and Related Disorders* 2006;**12**(2):87–92.
13. Morris ME, Huxham FE, McGinley J, Iansek R. Gait disorders and gait rehabilitation in Parkinson's disease. *Advances in Neurology* 2001;**87**:347–61.
14. Huxham F, Baker R, Morris ME, Iansek R. Head and trunk rotation during walking turns in Parkinson's disease. *Movement Disorders: Official Journal of the Movement Disorder Society* 2008;**23**(10):1391–7.
15. Visser JE, Voermans NC, Oude Nijhuis LB, et al. Quantification of trunk rotations during turning and walking in Parkinson's disease. *Clinical Neurophysiology: Official Journal of the International Federation of Clinical Neurophysiology* 2007;**118**(7):1602–6.
16. Herman T, Giladi N, Hausdorff JM. Properties of the 'timed up and go' test: more than meets the eye. *Gerontology* 2011;**57**(3):203–10.
17. King LA, Mancini M, Priest K, Salarian A, Rodrigues-de-Paula F, Horak F. Do clinical scales of balance reflect turning abnormalities in people with Parkinson's disease? *Journal of Neurologic Physical Therapy: Journal of Neurologic Physical Therapy* 2012;**36**(1):25–31.
18. Franzen E, Paquette C, Gurfinkel VS, Cordo PJ, Nutt JG, Horak FB. Reduced performance in balance, walking and turning tasks is associated with increased neck tone in Parkinson's disease. *Experimental Neurology* 2009;**219**(2):430–8.
19. Hong M, Perlmutter JS, Earhart GM. A kinematic and electromyographic analysis of turning in people with Parkinson disease. *Neurorehabilitation and Neural Repair* 2009;**23**(2):166–76.
20. Steiger MJ, Thompson PD, Marsden CD. Disordered axial movement in Parkinson's disease. *Journal of Neurology, Neurosurgery, and Psychiatry* 1996;**61**(6):645–8.
21. Orendurff MS, Segal AD, Berge JS, Flick KC, Spanier D, Klute GK. The kinematics and kinetics of turning: limb asymmetries associated with walking a circular path. *Gait and Posture* 2006;**23**(1):106–11.
22. Fino PC, Frames CW, Lockhart TE. Classifying step and spin turns using wireless gyroscopes and implications for fall risk assessments. *Sensors* 2015;**15**(5):10676–85.
23. Fino PC, Lockhart TE, Fino NF. Corner height influences center of mass kinematics and path trajectory during turning. *Journal of Biomechanics* 2015;**48**(1):104–12.

24. Mak MK, Patla A, Hui-Chan C. Sudden turn during walking is impaired in people with Parkinson's disease. *Experimental Brain Research* 2008;**190**(1):43–51.
25. Mellone S, Mancini M, King LA, Horak FB, Chiari L. The quality of turning in Parkinson's disease: a compensatory strategy to prevent postural instability? *Journal of Neuroengineering and Rehabilitation* 2016;**13**:39.
26. Mancini M, El-Gohary M, Pearson S, et al. Continuous monitoring of turning in Parkinson's disease: rehabilitation potential. *NeuroRehabilitation* 2015;**37**(1):3–10.
27. Bhatt H, Pieruccini-Faria F, Almeida QJ. Dynamics of turning sharpness influences freezing of gait in Parkinson's disease. *Parkinsonism and Related Disorders* 2013;**19**(2): 181–5.
28. Hase K, Stein RB. Turning strategies during human walking. *Journal of Neurophysiology* 1999;**81**(6):2914–22.
29. Song J, Sigward S, Fisher B, Salem GJ. Altered dynamic postural control during step turning in persons with early-stage Parkinson's disease. *Parkinson's Disease* 2012;**2012**: 386962.
30. Galna B, Lord S, Daud D, Archibald N, Burn D, Rochester L. Visual sampling during walking in people with Parkinson's disease and the influence of environment and dual-task. *Brain Research* 2012;**1473**:35–43.
31. Lohnes CA, Earhart GM. Saccadic eye movements are related to turning performance in Parkinson disease. *Journal of Parkinson's Disease* 2011;**1**(1):109–18.
32. Anastasopoulos D, Ziavra N, Savvidou E, Bain P, Bronstein AM. Altered eye-to-foot coordination in standing parkinsonian patients during large gaze and whole-body reorientations. *Movement Disorders: Official Journal of the Movement Disorder Society* 2011;**26**(12):2201–11.
33. Akram SB, Frank JS, Fraser J. Coordination of segments reorientation during on-the-spot turns in healthy older adults in eyes-open and eyes-closed conditions. *Gait and Posture* 2010;**32**(4):632–6.
34. Keus SH, Bloem BR, Hendriks EJ, Bredero-Cohen AB, Munneke M. Evidence-based analysis of physical therapy in Parkinson's disease with recommendations for practice and research. *Movement Disorders: Official Journal of the Movement Disorder Society* 2007; **22**(4):451–60. quiz 600.
35. Schenkman ML, Clark K, Xie T, Kuchibhatla M, Shinberg M, Ray L. Spinal movement and performance of a standing reach task in participants with and without Parkinson disease. *Physical Therapy* 2001;**81**(8):1400–11.
36. Vaugoyeau M, Viallet F, Aurenty R, Assaiante C, Mesure S, Massion J. Axial rotation in Parkinson's disease. *Journal of Neurology, Neurosurgery, and Psychiatry* 2006;**77**(7):815–21.
37. Morris ME. Movement disorders in people with Parkinson disease: a model for physical therapy. *Physical Therapy* 2000;**80**(6):578–97.
38. Hulbert S, Ashburn A, Robert L, Verheyden G. A narrative review of turning deficits in people with Parkinson's disease. *Disability and Rehabilitation* 2014:1–8.
39. Becker W, Nasios G, Raab S, Jurgens R. Fusion of vestibular and podokinesthetic information during self-turning towards instructed targets. *Experimental Brain Research* 2002; **144**(4):458–74.
40. Jurgens R, Boss T, Becker W. Estimation of self-turning in the dark: comparison between active and passive rotation. *Experimental Brain Research* 1999;**128**(4):491–504.
41. Jurgens R, Nasios G, Becker W. Vestibular, optokinetic, and cognitive contribution to the guidance of passive self-rotation toward instructed targets. *Experimental Brain Research* 2003;**151**(1):90–107.
42. Koutakis P, Mukherjee M, Vallabhajosula S, Blanke DJ, Stergiou N. Path integration: effect of curved path complexity and sensory system on blindfolded walking. *Gait and Posture* 2013;**37**(2):154–8.

43. Savona F, Stratulat AM, Roussarie V, Bourdin C. The influence of yaw motion on the perception of active vs passive visual curvilinear displacement. *Journal of Vestibular Research: Equilibrium and Orientation* 2015;**25**(3–4):125–41.
44. Earhart GM, Stevens ES, Perlmutter JS, Hong M. Perception of active and passive turning in Parkinson disease. *Neurorehabilitation and Neural Repair* 2007;**21**(2):116–22.
45. Zampieri C, Salarian A, Carlson-Kuhta P, Aminian K, Nutt JG, Horak FB. The instrumented timed up and go test: potential outcome measure for disease modifying therapies in Parkinson's disease. *Journal of Neurology, Neurosurgery, and Psychiatry* 2010;**81**(2):171–6.
46. Curtze C, Nutt JG, Carlson-Kuhta P, Mancini M, Horak FB. Objective gait and balance impairments relate to balance confidence and perceived mobility in people with Parkinson disease. *Physical Therapy* 2016;**96**(11):1734–43.
47. Bloem BR, Grimbergen YA, Cramer M, Willemsen M, Zwinderman AH. Prospective assessment of falls in Parkinson's disease. *Journal of Neurology* 2001;**248**(11):950–8.
48. Weaver TB, Robinovitch SN, Laing AC, Yang Y. Falls and Parkinson's disease: evidence from video recordings of actual fall events. *Journal of the American Geriatrics Society* 2016;**64**(1):96–101.
49. Robinovitch SN, Feldman F, Yang Y, et al. Video capture of the circumstances of falls in elderly people residing in long-term care: an observational study. *Lancet* 2013;**381**(9860):47–54.
50. Spildooren J, Vercruysse S, Heremans E, et al. Head-pelvis coupling is increased during turning in patients with Parkinson's disease and freezing of gait. *Movement Disorders: Official Journal of the Movement Disorder Society* 2013;**28**(5):619–25.
51. Spildooren J, Vinken C, Van Baekel L, Nieuwboer A. Turning problems and freezing of gait in Parkinson's disease: a systematic review and meta-analysis. *Disability and Rehabilitation* 2018:1–11.
52. Mancini M, Weiss A, Herman T, Hausdorff JM. Turn around freezing: community-living turning behavior in people with Parkinson's disease. *Frontiers in Neurology* 2018;**9**:18.
53. Connolly BS, Lang AE. Pharmacological treatment of Parkinson disease: a review. *Jama* 2014;**311**(16):1670–83.
54. Bohnen NI, Cham R. Postural control, gait, and dopamine functions in parkinsonian movement disorders. *Clinics in Geriatric Medicine* 2006;**22**(4):797–812 [vi].
55. Curtze C, Nutt JG, Carlson-Kuhta P, Mancini M, Horak FB. Levodopa is a double-edged sword for balance and gait in people with Parkinson's disease. *Movement Disorders: Official Journal of the Movement Disorder Society* 2015;**30**(10):1361–70.
56. Conradsson D, Paquette C, Lokk J, Franzen E. Pre- and unplanned walking turns in Parkinson's disease - effects of dopaminergic medication. *Neuroscience* 2017;**341**:18–26.
57. McNeely ME, Earhart GM. The effects of medication on turning in people with Parkinson disease with and without freezing of gait. *Journal of Parkinson's Disease* 2011;**1**(3):259–70.
58. Huxham F, Baker R, Morris ME, Iansek R. Footstep adjustments used to turn during walking in Parkinson's disease. *Movement Disorders: Official Journal of the Movement Disorder Society* 2008;**23**(6):817–23.
59. Horak FB, Dimitrova D, Nutt JG. Direction-specific postural instability in subjects with Parkinson's disease. *Experimental Neurology* 2005;**193**(2):504–21.
60. King LA, Horak FB. Lateral stepping for postural correction in Parkinson's disease. *Archives of Physical Medicine and Rehabilitation* 2008;**89**(3):492–9.
61. Lohnes CA, Earhart GM. Effect of subthalamic deep brain stimulation on turning kinematics and related saccadic eye movements in Parkinson disease. *Experimental Neurology* 2012;**236**(2):389–94.
62. St George RJ, Nutt JG, Burchiel KJ, Horak FB. A meta-regression of the long-term effects of deep brain stimulation on balance and gait in PD. *Neurology* 2010;**75**(14):1292–9.

63. Cheng FY, Yang YR, Wu YR, Cheng SJ, Wang RY. Effects of curved-walking training on curved-walking performance and freezing of gait in individuals with Parkinson's disease: a randomized controlled trial. *Parkinsonism and Related Disorders* 2017;**43**:20–6.
64. King LA, Salarian A, Mancini M, et al. Exploring outcome measures for exercise intervention in people with Parkinson's disease. *Parkinson's Disease* 2013;**2013**:572134.
65. Nieuwboer A, Baker K, Willems AM, et al. The short-term effects of different cueing modalities on turn speed in people with Parkinson's disease. *Neurorehabilitation and Neural Repair* 2009;**23**(8):831–6.
66. Willems AM, Nieuwboer A, Chavret F, et al. Turning in Parkinson's disease patients and controls: the effect of auditory cues. *Movement Disorders: Official Journal of the Movement Disorder Society* 2007;**22**(13):1871–8.

CHAPTER

8

Is freezing of gait a balance disorder?

Clinical case

Seven years after his diagnosis of Parkinson disease (PD), Stanley noticed slight "hesitation" when initiating gait, that is, a significant delay in taking the first step after intending to walk. He also noticed the most freezing during a turn and when performing other activities while walking or anxious. Eventually, the duration of these freezing episodes resulted in falls onto his knees and hands, which increased anxiety even more. His freezing improved much when he was On but freezing became a problem when he was in the Off state during motor fluctuations. Eventually, he relied on external cues, such as a laser pointer and metronome to help him overcome freezing but is still severely impaired by his freezing and associated falls.

A. What is freezing of gait and why it is associated with falls

Freezing of gait (FoG) is a unique and disabling clinical phenomenon characterized by brief episodes of inability to step or step by extremely short steps.[1] FoG is triggered by postural transitions (change in postural set) such as initiating gait, turning while walking, narrow passages, obstacle crossing, or approaching a destination.[1] Freezing episodes are often described as if "the feet are glued to the floor" while the body center of mass continues to move forward, sometimes resulting in a forward fall.[2] Since sudden, unexpected freezing can result in imbalance with reduced ability to take a step to recover equilibrium, FoG is one the most common reasons for falls in people with PD.[3]

Balance Dysfunction in Parkinson's Disease
https://doi.org/10.1016/B978-0-12-813874-8.00008-8

FoG is very common in PD and increases in prevalence the longer the duration of PD. A long-term prospective follow-up of 136 patients with newly diagnosed PD showed a high incidence of FoG (81%) at 20 years.[4] Another prospective study revealed a high prevalence of FoG (87%) at a follow-up of 11 years.[5] FoG can also be experienced in a relatively early stage of PD and even in untreated patients, but usually starts very mild in these cases.[6] The DATATOP study found that an absence of tremor, the presence of a gait disorder, and the development of balance and speech problems are associated with the occurrence of FoG.[6] FoG is usually observed in the Off levodopa state.[7]

FoG is also one of the leading predictors of falls, despite the fact that FoG normally occurs when people are Off and falls usually occur when people are in On.[8,9] It is likely that freezers are more likely to fall because they have more severe balance and gait disorders, even when they don't freeze. A study of 205 community-dwelling people with PD, without cognitive impairment, showed that three simple clinical observations accurately predicted falls: falling in the previous year, FoG in the past month, and slow gait.[3]

It is important to understand that FoG is usually not associated with lack of motion like a frozen statue. More common is "trembling knees," consisting of fast oscillations of the legs at a frequency spectrum of 2–6 Hz with the patient's attempts to walk.[10–14] Moore[11] found that trembling of the knees characterized 89% of spontaneous FoG episodes in a group of people with PD. Schaafsma[14] also found 84% of their population of PD patients with FoG had "trembling in place." Fig. 8.1 shows angular displacement of the right and left knee during walking and a higher frequency, lower amplitude trembling of the knees during a freezing episode in a patient with PD. This patient also shows a

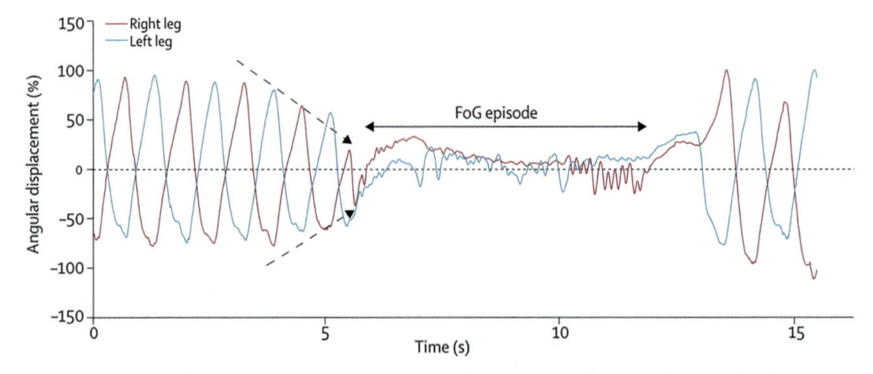

FIGURE 8.1 FoG episode as measured with angular displacement of the knees (% maximum knee angle) in a representative subject with FoG. *From Nutt JG, Bloem BR, Giladi N, Hallett M, Horak FB, Nieuwboer A. Freezing of gait: moving forward on a mysterious clinical phenomenon.* The Lancet Neurology *2011;**10**(8):734–44.*

shortening of the steps (sometime associated with quicker stepping called hastening) just prior to a freezing episode.

Freezing may not be limited to gait as some patients with FoG also show "freezing," or motor blocks, motor arrests, hesitation, or movement breakdown during repetitive upper or lower extremity or speech motions.[15,16] Motor tasks involving bilateral rhythm and amplitude control such as pedaling, stepping in place, alternating foot motion, finger or hand tapping, hand drawing, hand pronation-supination, reading out loud have been shown to be associated with such motor blocks.[15,16] Fig. 8.2 shows an example of freezing during a bilateral finger

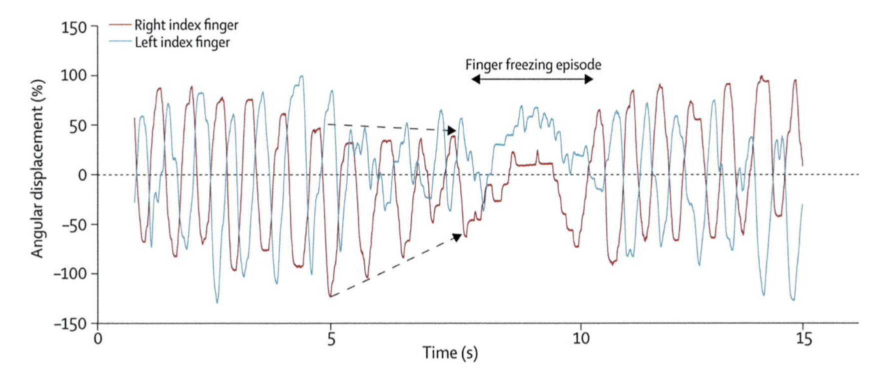

FIGURE 8.2 Example of "freezing" episode in the arm of a subject with PD during rhythmical motions. *From Nutt JG, Bloem BR, Giladi N, Hallett M, Horak FB, Nieuwboer A. Freezing of gait: moving forward on a mysterious clinical phenomenon.* The Lancet Neurology *2011;**10**(8): 734–44.*

tapping task.[1] Based on visual inspection, these nongait freezing events occur more likely in people who report having FoG (57%–73%) than those without FoG (0%–8%), suggesting some common underlying mechanisms. However, these nongait freezing events are much less common than FoG, perhaps because freezing during gait involves abnormal coupling between two types of motor control, posture and gait.

FoG has well-known links to mental function. It is precipitated by visual spatial perceptions such as narrow doorways or elevator doors, worse with distractions and dual-tasks and stress or anxiety. FoG is also ameliorated with intense emotions, paying attention to voluntary aspects of gait and external cues, such as auditory cues, lines on the ground, or stairs. Several, but not all, studies show worse executive function, particularly executive response inhibition in freezers compared to nonfreezers, albeit not exclusively. For example, Fig. 8.3 shows differences in three tests of response inhibition, Stroop conflict time, Stroop interference score, and Go-NoGo target misses in which more freezers (FR) than nonfreezers (NF) show poor performance compared to healthy controls

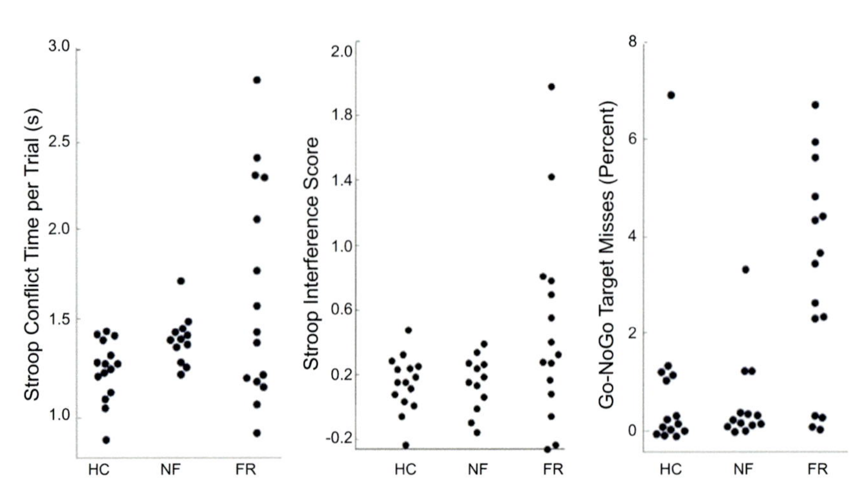

FIGURE 8.3 Worse cognitive inhibition in a group of PD with FoG (FR) compared to healthy controls (HC) or nonfreezers with PD (NF). *From Cohen RG, Klein KA, Nomura M, et al. Inhibition, executive function, and freezing of gait.* Journal of Parkinson's Disease *2014; 4(1):111–22.*

(HC).[17] These subjects were tested Off. Other studies do not find such excessive impairments in executive function when tested On (Morris, Nutt, and Horak, submitted).

B. Do freezers have more balance disorders than nonfreezers?

PD with FoG appears to show more balance disorders than nonfreezers. However, it is important to control for severity of disease in these studies, as when FoG becomes more likely the more severe the motor and cognitive signs, which also are associated with balance disorders. A recent systematic review showed that people with FoG have worse performance on clinical balance tests, such as the Mini-BEST, Berg Balance Scale, etc., than PD without freezing (see Meta-analysis summary of clinical balance tests in Fig. 8.4).[18] However, it is important to understand which specific balance domains are most affected in those with FoG to focus on balance rehabilitation, rehabilitation aimed to reduce FoG.

Anticipatory postural adjustments

Slow, long duration anticipatory postural adjustments (APAs) include abnormal pauses that disrupt posture-movement coordination and may precipitate freezing of gait (FoG). However, an APA is almost always present during voluntary step initiation, except in some people with PD who are akinetic. Thus, freezing does not appear to be due to absent APAs.

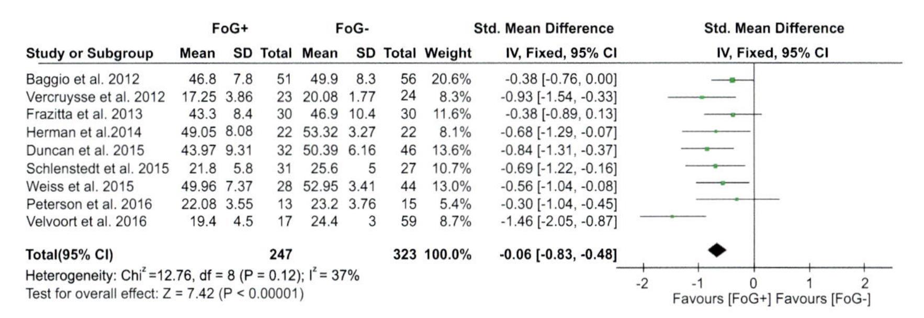

Study or Subgroup	FoG+ Mean	FoG+ SD	FoG+ Total	FoG- Mean	FoG- SD	FoG- Total	Weight	Std. Mean Difference IV, Fixed, 95% CI
Baggio et al. 2012	46.8	7.8	51	49.9	8.3	56	20.6%	-0.38 [-0.76, 0.00]
Vercruysse et al. 2012	17.25	3.86	23	20.08	1.77	24	8.3%	-0.93 [-1.54, -0.33]
Frazitta et al. 2013	43.3	8.4	30	46.9	10.4	30	11.6%	-0.38 [-0.89, 0.13]
Herman et al.2014	49.05	8.08	22	53.32	3.27	22	8.1%	-0.68 [-1.29, -0.07]
Duncan et al. 2015	43.97	9.31	32	50.39	6.16	46	13.6%	-0.84 [-1.31, -0.37]
Schlenstedt et al. 2015	21.8	5.8	31	25.6	5	27	10.7%	-0.69 [-1.22, -0.16]
Weiss et al. 2015	49.96	7.37	28	52.95	3.41	44	13.0%	-0.56 [-1.04, -0.08]
Peterson et al. 2016	22.08	3.55	13	23.2	3.76	15	5.4%	-0.30 [-1.04, -0.45]
Velvoort et al. 2016	19.4	4.5	17	24.4	3	59	8.7%	-1.46 [-2.05, -0.87]
Total(95% CI)			**247**			**323**	**100.0%**	**-0.06 [-0.83, -0.48]**

Heterogeneity: Chi² =12.76, df = 8 (P = 0.12); I² = 37%
Test for overall effect: Z = 7.42 (P < 0.00001)

FIGURE 8.4 Meta-analysis showing worse clinical balance scale performance in freezers compared to nonfreezers with PD. *From Bekkers EMJ, Dijkstra BW, Dockx K, Heremans E, Verschueren SMP, Nieuwboer A. Clinical balance scales indicate worse postural control in people with Parkinson's disease who exhibit freezing of gait compared to those who do not: a meta-analysis.* Gait and Posture *2017;**56**:134–40.*

In fact, Schlenstedt and colleagues[19] showed that people with PD who freeze do not show smaller APA amplitude (Fig. 8.5A) or longer duration prior to a step compared to PD who do not freeze. During dual-tasks, however, freezers did reduce the size of their APAs, unlike nonfreezers, consistent with less automatic step initiation in freezers (Fig. 8.5B). Interestingly, a significant association between the size of APA and the freezing of gait questionnaire was reported indicating that the more severe the perceived FoG, the larger (not smaller) the size of medio-lateral (ML) APAs (Fig. 8.5C).

One explanation for the reduced size of APAs when freezers do not show a freezing episode might be that the reduction of size of ML APAs is a compensatory strategy to successfully initiate gait in subjects with PD who experience FoG. The compensatory, small APAs might be needed as a result of posture-gait coupling deficits in subjects with FoG. Subjects with PD who freeze might not be able to control large accelerations of the body center of mass, which might cause a failure of coupling the APA with the stepping pattern. Perhaps it is difficult for freezers to inhibit the postural set and initiate locomotion. The abnormally large APA that occurs only during trials with freezing events could indicate a total breakdown of the usual gait initiation pattern and its compensatory reduced APA size, resulting in a freezing episode.

Automatic postural responses

The first paper to suggest that compensatory stepping responses to external perturbations were abnormal in people with PD and freezing concluded that "trembling of knees" during FoG episodes could represent multiple APAs.[20] Ten subjects with severe FoG were tested both Off and On. Results showed that 8/10 subjects in the Off state showed several alternating right–left leg, loading-unloading cycles prior to a delayed

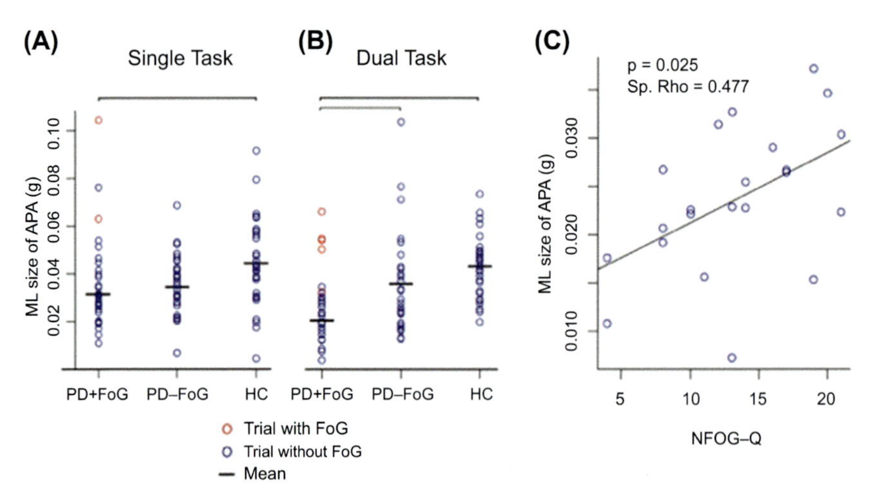

FIGURE 8.5 A,B Freezers show smaller APAs than nonfreezers during dual-tasking. Exceptions of trials with larger than normal APAs are during the trials with an actual FoG event (*red*). C Relationship between size of APA and score on new FoG questionnaire. *From Schlenstedt C, Mancini M, Nutt J, et al. Are hypometric anticipatory postural adjustments contributing to freezing of gait in Parkinson's disease?* Frontiers in Aging Neuroscience *2018;**10**:36.*

compensatory step or a "fall" without a step. Fig. 8.6 shows the reciprocal activation of tibialis and gastrocnemius, as well as between left and right tensor facia latae muscles that accompanied the repetitive left-right weight shifting prior to a compensatory step in a freezer. These multiple APAs oscillate on average at 2.67 Hz, within the knee trembling range. When more than one APA is sequenced together, the amplitudes of the APAs gradually get larger until they are larger than normal before a delayed, or absent, step. The more the APAs were repeated, the longer the latency until the step was initiated. The shorter than normal latency to the first weight shift and large amplitude of multiple APAs in freezers, suggest that FoG during forward disequilibrium is caused by abnormal coupling between the APA and the step motor programs rather than by an impaired ability to generate an APA. Of course, the mechanism for freezing in response to surface displacements causing forward imbalance with the body center of mass (CoM) in front of the feet may differ than when freezing occurs when people are upright, prior to a voluntary step.

A more recent study showed that the step length of postural stepping responses were smaller in freezers than nonfreezers.[21] In addition, people who freeze did not improve or retain improvement in protective postural responses as well as nonfreezers with repetitive practice of responding to surface translations.[22] Specifically, CoM displacement after perturbations did not change significantly in freezers compared to nonfreezers with practice over 35 trials.[22]

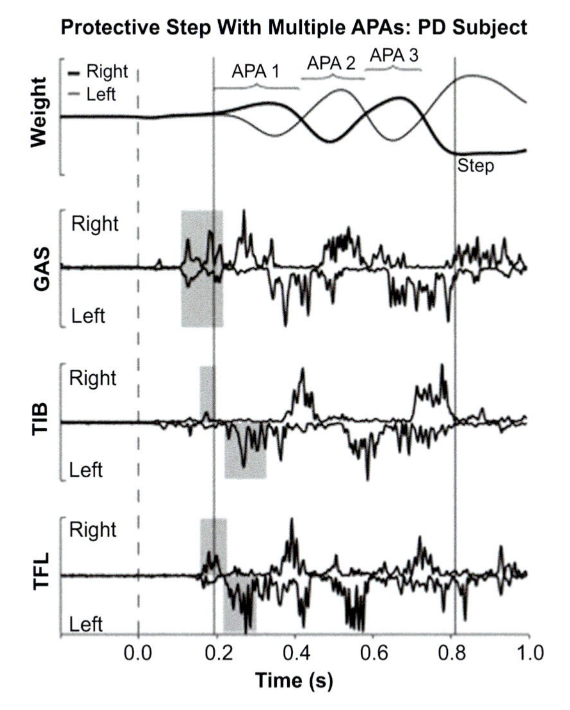

FIGURE 8.6 Abnormal reciprocal EMG pattern of the leg antagonists during "trembling of the legs." The shaded areas show EMG bursts that were quantified in the study prior to or during the first APA. *From Jacobs JV, Nutt JG, Carlson-Kuhta P, Stephens M, Horak FB. Knee trembling during freezing of gait represents multiple anticipatory postural adjustments.* Experimental Neurology *2009;**215**(2):334–41.*

We recently instrumented the Push and Release test of postural responses with bodyworn inertial sensors that can quantify the latency to the first step, length of the first step, number of steps, and time to achieve equilibrium, that is time to stop backward displacement after a fall starts.[23,24] We found normal, or shorter than normal, latencies of stepping responses in people with PD than control subjects. These short latencies demonstrate that postural response latencies are not delayed by bradykinesia and are often too early, as if they release a response without adequate planning.[25] As expected, we also found significantly shorter and narrower compensatory steps, more steps, and longer time to equilibrium from 38 PD patients with objectively confirmed FoG compared to 44 PD patients without FoG or 24 elderly control subjects (unpublished results). Our PD groups had similar severity based on Motor UPDRS scores such that differences in compensatory stepping responses were more likely due to freezing than general bradykinesia, affecting externally triggered postural responses.

Sensory orientation for standing posture

It is not easy to predict who will have FoG based simply on a clinical observation of standing postural alignment, body sway, or limits of stability. However, the body CoM has been measured as too far forward in early PD but too far backward in later PD,[26] in the patients who are more likely to have FoG. It has been hypothesized that freezing is associated with more flexed posture, pressure on the forefoot, forward head, and increased axial tone but studies are needed to test the contribution of these types of musculoskeletal changes to FoG.

Regarding postural sway, five out of eight studies found no differences in postural sway during quiet standing with eyes open in freezers versus nonfreezers.[27–29] However, the largest studies report a backward shift of the body CoP[30,31] and increased anterior-posterior root mean square of sway in freezers compared to nonfreezers standing with eyes open.[32] Nevertheless, sway increased in velocity and area with eyes closed compared to eyes open, similar to controls, in both freezers and nonfreezers with no group differences.[32] We are finding that voluntary limits of stability are smaller in freezers than nonfreezers (unpublished results), even when correcting for disease duration. Smaller limits of stability could reflect poorer proprioception/kinesthesia, greater fear of falling, poorer flexed postural alignment, or weaker strength in freezers.

It is not clear whether sensory weighting to control postural sway in standing is different in freezers than nonfreezers, or even between people with PD versus age-matched control subjects. Few small studies have inconsistent findings, likely because they did not control for severity of disease, the difficulty in identifying freezers and test subjects in different medication states. For example, one study found increased postural sway in freezers compared to nonfreezers in the Off state with eyes closed[30] and another one increased sway in the On state when standing.[31] Our recent study of quantitative sensory weighting when surface and/or visual inputs change based on a quantitative model of sensory integration[33] found few differences between PD and age-matched control subjects, except less increase in visual/vestibular weighting and decrease in somatosensory sensory weighting when the surface rotations increased from one to two degrees.[33a] In contrast, another study found abnormal sensory weighting but freezers were not compared to nonfreezers in these studies.[34] Regardless, standing posture seems to be the least affected balance domain in people with FoG, compared to nonFoG.

Dynamic postural stability during gait

Variability and coordination of stepping during walking, even when no freezing occurs, have been shown to be worse in PD with FoG than without FoG.[35–39] For example, FoG can sometimes, but not always, be associated with progressively smaller hypokinetic steps (the sequence effect) and faster (hastening effect) immediately prior to freezing.[40] Left-right timing such as stride time variability and the phase coordination index can also be more abnormal in freezers than nonfreezers, even when they do not display a freezing event.[35–39]

Dual-task cost on gait is also larger in PD with FoG than in PD without FoG. Specifically, stride length and first-step duration were more affected by serial subtraction in freezers than nonfreezers, even when controlling for severity of disease.[32] Fig. 8.7 compares single- and dual-task stride length and the anticipatory postural adjustment first-step duration in nonfreezers (FoG−), freezers (FoG+), and age-matched controls (*dark and light gray lines*). Dual-task cost in freezers was correlated with their cognitive deficits, specifically, ability to release a response in a Go-No-Go task and simple reaction time.[41] The greater variability of gait, as well as the larger dual-task cost on gait, suggests that gait is less automatic, that is more driven from higher cortical areas, in PD with FoG than PD without FoG.

Could FoG be due to the loss of automatic control of walking? The notion is that, unlike healthy subjects who use more automatic basal ganglia-cerebellar-brainstem circuits (with SMA) to control gait, freezers depend more upon frontal cortical control of STN and PPN for a top-down, less automatic control (see Chapter 6, Fig. 6.6). Levodopa helps the return to more automaticity in control of balance and gait. This may

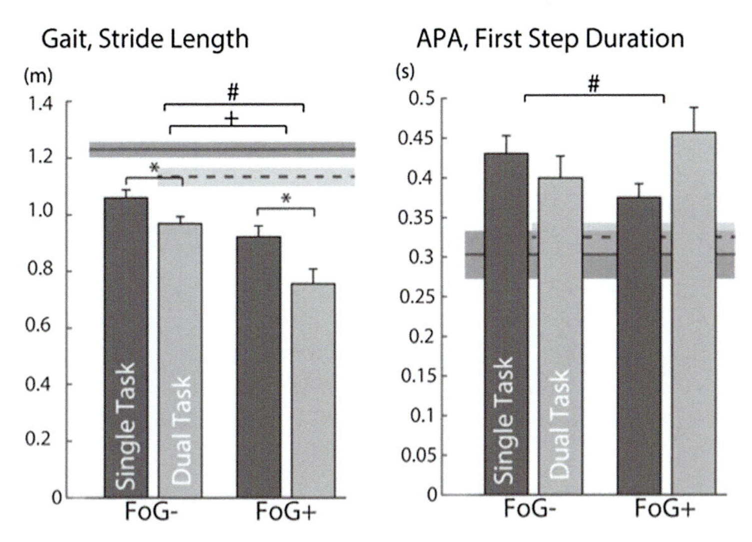

FIGURE 8.7 Freezers show larger dual-task cost on stride length and first-step duration than nonfreezers. *From de Souza Fortaleza AC, Mancini M, Carlson-Kuhta P, et al. Dual task interference on postural sway, postural transitions and gait in people with Parkinson's disease and freezing of gait.* Gait and Posture *2017;**56**:76–81.*

occur because dopaminergic depletions first affect the posterior putamen and caudate, resulting in primarily motor symptoms. The cognitive striatum is comparatively spared and forces patients into an increased reliance on the attention-controlled, less skilled behavior.[15] Loss of dopamine also leads to increased crossover of nonmotor information in generation of motor loops through the basal ganglia, worsening dual-task cost. This creates a system at risk for temporary jamming of central motor-cognitive information processing.[42]

Consistent with the suggestion that gait is less automatic in freezers, recent studies showed enhanced activation of the frontal cortex during walking using functional near-infrared Spectroscopy (fNIRS) while walking and functional MRI (fMRI) while in a scanner imagining walking. Healthy subjects use more frontal lobe activity at the beginning of gait and immediately prior to turning, with low frontal activity during continued walking and turning itself (Ref. 43 and unpublished data). Freezers were also slower to reduce frontal activity during continued walking and increased frontal activity even more during turning when freezing.[43] Similarly, a virtual reality paradigm of lower limb freezing in an fMRI scanner showed increased frontoparietal activity and reduced sensorimotor and basal ganglia activity during freezing episodes.[42,54]. An fMRI study of freezing of repetitive finger movements showed similar increased activation of frontal cortical areas when freezing occurred.[15] Fig. 8.8 shows the similarities in the increased frontal cortical but

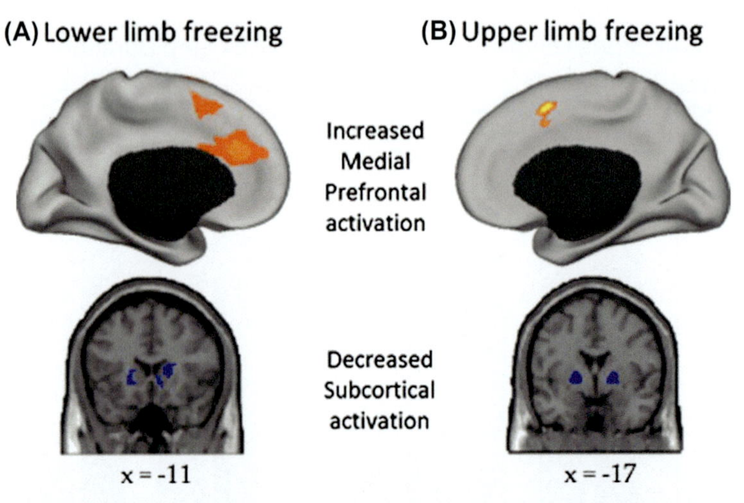

FIGURE 8.8 Overlap in the brain activation between freezing and nonfreezing cyclical movement of the (A) lower limbs and (B) upper limbs. Both types of freezing showed dissociation between associative, frontal cortical areas and the basal ganglia. *From Vercruysse S, Gilat M, Shine JM, Heremans E, Lewis S, Nieuwboer A. Freezing beyond gait in Parkinson's disease: a review of current neurobehavioral evidence.* Neuroscience and Biobehavioral Reviews *2014;***43***: 213–27.*

decreased subcortical activity during nongait, lower limb, and upper limb freezing-like events during virtual reality in a scanner.[15]

One of the most reliable way to observe FoG in the clinic is to ask the patient to attempt to turn 360 degrees in place to the right and then to the left.[43a] Although it is not known why turning, in particular, induces FoG, it may involve the tight coupling between anticipatory postural adjustments and top-down modification of the stepping pattern to maintain stability and change direction. Although turn duration and number of steps needed to turn can be used to measure freezing during turning, these measures are less sensitive than using a quantitative FoG ratio, based on trembling of the knees.[44] The FoG ratio is calculated from the power spectrum as the area under the curve for trembling, freezing frequencies (3.5–8 Hz) divided by the area under the curve for the stepping frequencies (0.5–3 Hz).[44] The FoG ratio is also correlated with the New Freezing of Gait Questionnaire,[45] NFOG-Q, and by expert neurologists using video rating. Fig. 8.9 shows

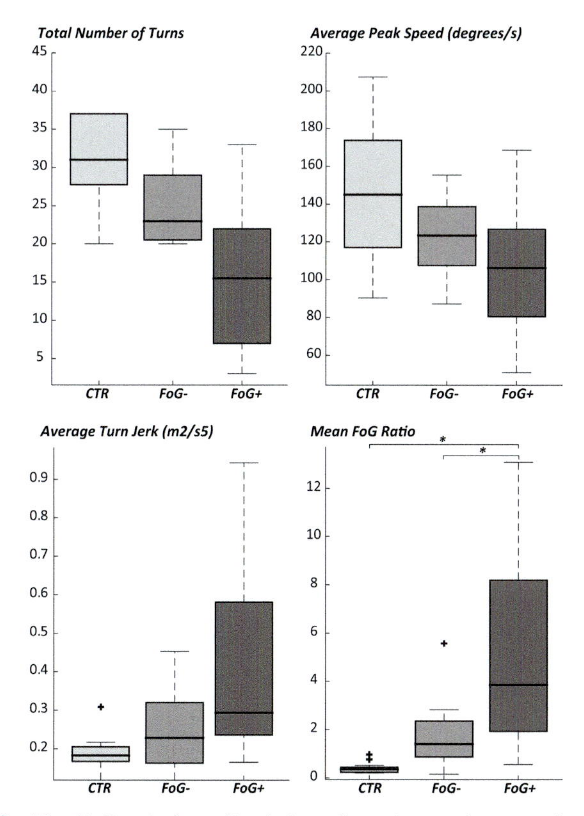

FIGURE 8.9 The FoG ratio best discriminated turning performance between freezers (FoG+) and nonfreezers (FoG−) or controls (CTR), compared to number of turns, peak turn speed, and jerkiness of the trunk during turning. *From Mancini M, Smulders K, Cohen RG, Horak FB, Giladi N, Nutt JG. The clinical significance of freezing while turning in Parkinson's disease.* Neuroscience *2017;***343***:222–8.*

significant difference in the FoG ratio during turning with 16 freezers and 12 nonfreezers but no significant differences (just trends) for the other measures of turning.[44]

C. Do brain circuitry abnormalities in freezers suggest causes for FoG?

FoG has been associated with both structural and functional changes in multiple brain networks, both motor and nonmotor. Previous studies have suggested possible alterations in the frontal lobe-basal ganglia-cerebellar-brainstem network that controls initiation and maintenance of gait.[46] The cerebellum has also been regarded as possibly important contributor to FoG since isolated lesions within the locomotor cerebellar areas can cause FoG.[47] Imaging studies most consistently reveal altered connectivity between the sensorimotor[48,49] and frontoparietal and visual networks (as well as with subcortical areas) in freezers compared to nonfreezers.[50,51]

One hypothesis that has support in the literature is that brain networks that control nonmotor, cognitive processes, specifically response inhibition and/or attention, are abnormal in people with FoG compared to nonfreezers.[46] For example, Fig. 8.10 shows that people with FoG have increased functional, but decreased structural, connectivity between the right supplementary motor cortex (SMA) and the right pedunculopontine nucleus (PPN), compared to people without FoG or age-matched control

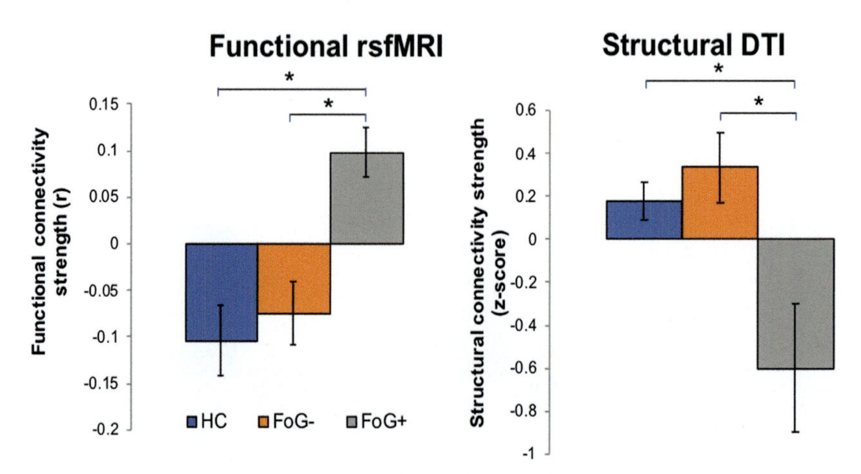

FIGURE 8.10 Freezers showed increased SMA-PPN functional connectivity but decreased structural connectivity (diffusion tensor imaging, DTI) compared to nonfreezers (FoG−) or controls (HC).

subjects.[48,52] In fact, the stronger the functional connectivity between the right SMA and PPN, the worse the FoG, consistent with a maladaptive compensation for loss of network structure.[48]

This right-hemisphere pathway is part of both the cognitive inhibition network, such that it is active when people do cognitive inhibition tasks.[49] The same pathway likely is also part of the locomotor network, consistent with greater executive inhibition deficits in people with FoG compared to those without FoG.[17,48,52,53] In addition, functional connectivity strength (Fig. 8.10, *left panel*) was significantly correlated with the NFOG-Q, the Objective FoG Index, and Clinical FoG expert rating, supporting the hypothesis that increasing functional connectivity between the SMA and locomotor areas in the brainstem (PPN) and STN in the basal ganglia is a maladaptive compensatory strategy for FoG.[48] Perhaps stronger control between the frontal cortex and subcortical and brainstem locomotor areas reflect less automatic control of balance and gait in freezers, compared to nonfreezers.

FoG appears to be specifically associated with abnormal attention network activity. Tessitore et al.[51] found that patients with FoG exhibit a significantly reduced functional connectivity only within the executive-attention and visual networks which was correlated with the FoG clinical severity. Shine et al.[42,54] then reported reduced activation in the ventral attention network, of which the insular cortex is an important component. They hypothesize that the anterior, dopamine-depleted striatum in PD becomes impaired in its ability to concurrently process information from segregated, yet complementary motor, cognitive, and limbic corticostriatal loops.[55] During certain challenging situations, the processing capacity of the striatum becomes overwhelmed, resulting in overactivation of the output nuclei of the basal ganglia that send excessive GABAergic inhibitory projections to thalamic and brainstem locomotor regions causing a breakdown of gait, and ultimately freezing.[55,56] According to this model, any increase in processing demands in the striatum would increase freezing. However, evidence is not strong that dual-tasking increases the severity or incidence of freezing.

Another hypothesis that has support of functional connectivity studies is that freezers have abnormal emotional brain circuitry that may explain why anxiety and panic attacks are often associated with freezing episodes.[57] FoG often coincides with panic attacks in PD.[57–61] For example, threatening situations (i.e., walking in the dark or crossing an elevated plank in virtual reality) provoke greater anxiety, and also elicit a greater amount of FoG in PD.[59,62] Visceral responses associated with anxiety, such as changes in heart rate and skin conductance, have also been shown to occur immediately prior to and during FoG episodes, although it is not clear if they are a cause or effect of FoG.[63,64] Consistent with the hypothesis that abnormal mood circuitry contributes to FoG is the increased resting state functional connectivity between the right amygdala and putamen and increased anticoupling between the left amygdala and the frontal-parietal network.[57] Furthermore, increased objective FoG severity (FoG ratio) was significantly associated with increased functional connectivity between the left amygdala to left putamen and increased anticoupling between the left amygdala to left lateral prefrontalcortex, left amygdala to right dorsolateral prefrontal cortex, and right amygdala to right intraparietal cortex.[57] In addition, in this study, fear of falling scores were negatively associated with the functional connectivity between both the left and right amygdala and the frontal-parietal network. Together, these results support the hypotheses that FoG is associated with increased baseline striatolimbic connectivity, in which FoG is likely exacerbated due to a lack of top-down control by the frontoparietal, attention network over the amygdala.[57]

In support of an emotional network cause of FoG, a study of 221 patients with PD found that the Freezing of Gait Questionnaire and the anxiety section of the Hospital Anxiety and Depression Scale were the strongest predictors and, alone, could significantly predict who would develop freezing of gait in the next 15 months with 82% accuracy.[65] The theory is that the frontoparietal network is often recruited to exert top-down control over the emotional responses from the threat system during instances of false alarm, or once the threat no longer poses a significant risk.[66–69] Freezers may not be able to control the amygdala response to threat (such as fear of falling) because of impaired frontal-parietal network,[42,50,51] accompanied by impaired executive function.[42,70–72] Freezers are known to rely on attentional resources to operate their gait due to a loss of automaticity[73] which could make them particularly susceptible for limbic interference, since both limbic and motor processes compete for attentional resources during gait.

Another hypothesis for FoG is reduced interlimb coordination mediated by the bilateral somatosensory cortex. The study by Lenka, 2016,[74] found reduced interhemispheric connectivity between the left parietal opercular cortex and the primary somatosensory and auditory cortical areas. They also observed correlations between these networks and FoG questionnaire scores. The parietal operculum may be important as this region corresponds to the secondary somatosensory area (S2) that could be responsible for poor kinesthesia in PD with FoG.[74]

Taken together, imaging studies suggest that higher level nonmotor, as well as sensorimotor cortical areas, and their connections to subcortical areas are involved in FoG. However, comparisons of brain function or structure between freezers and nonfreezers require controlling for severity of disease since FoG is more common in the advanced stages of the disease, when cognitive impairment is also more severe. In addition,

FoG may not be a bimodal distribution but rather a continuous function with gradually increasing freezing problems until they become more and more apparent and troublesome. As an episodic event, FoG is often difficult to elicit in the clinic or laboratory, so quantitative biomarkers, preferably with continuous metrics could be helpful.

D. How can FoG be treated with medication?

The most effective medication to reduce FoG is levodopa. In fact, over 90% of people with idiopathic PD who have FoG show significant improvement in freezing with levodopa and FoG limited to the On levodopa state is quite rare.[7] Monoamine oxidase B (MAO-B) inhibitors have been associated with decreased prevalence of FoG but treatment with these drugs is ineffective against FoG. Intravenous amantadine has been reported to decrease FoG but larger blinded studies have not confirmed this finding. L-Dihydroxyphenylserine (L-DOPS), the precursor for norepinephrine, is marketed for FoG in Japan but approval for this indication was based on a large open trial with L-DOPS. Its utility awaits a randomized controlled clinical trial. Methylphenidate has been reported to be of some assistance in FoG in patients with STN DBS[75] but another study of methylphenidate in gait disorders[76] found no effect of the drug in patients with gait disorders. Controlled clinical trials[77,78] of botulinum toxin injections in the gastrocnemius muscles did not confirm the improvement in FoG reported in open trials.[79]

E. Does deep brain stimulation improve FoG?

After DBS in STN, freezing can be reduced in about one-third of patients but also aggravated, with more falls, in a significant number of patients.[80] After surgery, FoG severity was proportional to FoG severity before surgery without dopaminergic treatment, as well as dependent upon the dopaminergic treatment dosage and severity of motor fluctuations after surgery.[80] After DBS in GPi, hypokinetic gait disorder and FoG is often worsened.[81] The gait disorder seen after DBS in GPi included shuffling steps and difficulties with gait initiation and turning, suggestive of FoG. Increasing DBS voltages improved dystonia but triggered more FoG, sometimes worsening over a period of a few hours but vanishing within minutes after ceasing DBS.[81] A recent review of clinical studies related to DBS for FoG in PD patients over the past decade suggested that DBS of either subthalamic nucleus (STN) or pedunculopontine nucleus (PPN) alone or in combination can improve the symptoms of FoG, although not in a majority of patients.[82] They conclude that prospective clinical trials with a larger sample size are needed to systematically assess the efficacy of DBS in any target for FoG.

Although the reasons are unclear, lower frequencies of DBS may be more beneficial for gait and for FoG, specifically, than the more common, high frequency stimulation used for tremor and bradykinesia of voluntary movement (see Chapter 6). A recent review of the literature found that lower frequencies of stimulation (<100 Hz) induced greater response for akinesia, gait, and FoG, including the number of freezing of gait events, compared to higher frequencies (>100 Hz).[83] It is likely that the more that DBS reduces motor fluctuations, the less time subjects suffer in the Off levodopa state, so the more efficacious is the DBS to reduce FoG (since it usually only occurs in the Off state).

An important limitation of all intervention studies for FoG is that it is difficult to quantify or elicit freezing for a study, so questionnaires are usually used as outcomes. In our experience, it is not uncommon for people who say they have no freezing to reveal freezing events in the laboratory and vice versa, so questionnaires are not ideal. Furthermore, we do believe freezing is a continuum, not a dichotomy among people with PD so future studies using wearable technology to quantify the quantity and severity of freezing events during daily life would lend great insight.

F. Does rehabilitation improve FoG?

Cueing has been shown to be helpful for temporarily overcoming prolonged FoG events, but not for reducing the frequency of FoG events.[84] Recently, various devices have been developed to provide visual cues (laser shoes[85,86]) or auditory cues (Gait Tutor[87]) or tactile cues (VibroGait[88,89]) to aid patients overcome freezing events. For example, Fig. 8.11 shows how freezing during 360-degree turning in place can be reduced with either an auditory metronome or with closed-loop tactile feedback on the wrists, driven by foot placement on the ground. Freezing in this study was quantified with the FoG ratio, as mentioned above.

However, these cueing tools have primarily been shown to be useful in prescribed walking tests in the laboratory and have not been evaluated for feasibility in real-life environments, apart from a small study from Barthel et al. showing preliminary but interesting results on improving the NFOG-Q after 1 week of using the laser shoes at home.[86] One of the problems with cues to overcome freezing is habituation, the gradual decrease in effectiveness of an external cue upon repetition. That is why many people with FoG tend to use multiple types of sensory cues to overcome their freezing episodes, such as laser lights, marching music, and light touch.

A handful of small rehabilitation intervention studies for FoG have reported mixed results. A small study on a bike-walking device reported

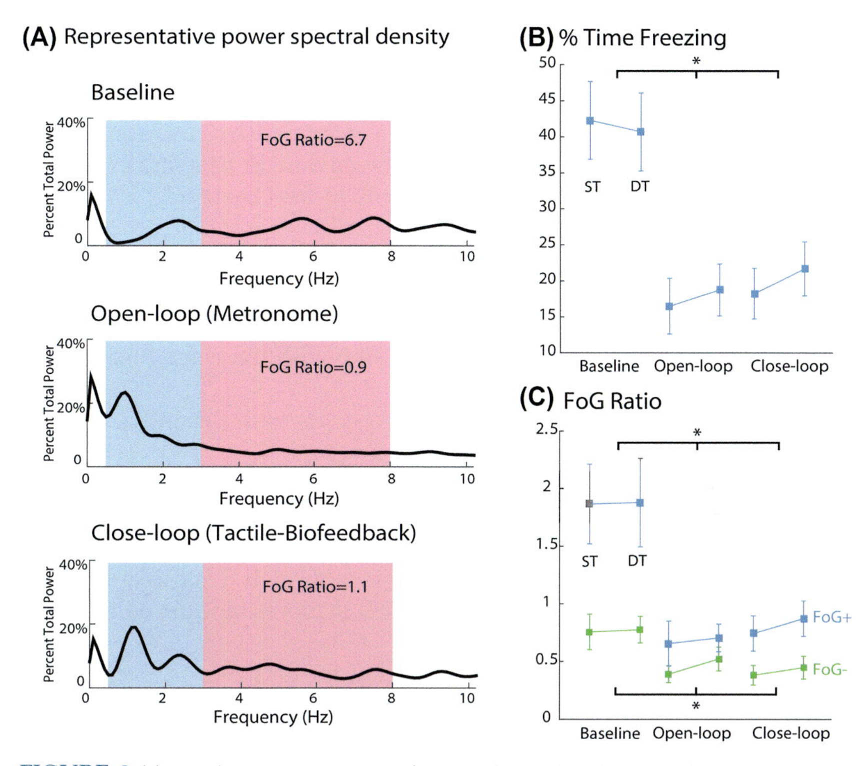

FIGURE 8.11 Reduction in Freezing of Gait Index with either open-loop external cues (metronome) or closed-loop cues (VibroGait). (A) Representative power spectral densities of the AP acceleration of the shins for baseline, open-loop, and closed-loop conditions together with the relative FoG ratio for that example. *Right panel*, freezing severity across the 3 conditions, measured by the (B) % time spent freezing (only in freezers) and (C) the FoG ratio (freezers and nonfreezers). *From Mancini M, Smulders K, Harker G, Stuart S, Nutt JG. Assessment of the ability of open- and closed-loop cueing to improve turning and freezing in people with Parkinson's disease.* Scientific Reports *2018;8(1):12773.*

reduced FoG in some, but not all, people with FoG.[90] Likewise, a small study of 20 people with FoG practicing dual-task on a virtual-reality device while stepping in place on a balance platform reported some improvements in dual-task performance and in gait measures, but not on the NFOG-Q.[91] Though there were some positive studies on effects of dance on balance in people with FoG,[92,93] the change in FoG, itself, with dance was not significant in a recent metaanalysis on the topic.[94] A larger, well-controlled, virtual-reality study of obstacle avoidance while walking on a treadmill showed significant improvements in gait, dual-task costs, and certain aspects of executive function, but not on the NFOG-Q.[95] Recently, we also completed a study of a 6-week Agility Boot Camp Intervention in 91 subjects with PD, half with FoG. We also found significant improvements in dual-task cost but no improvement in FoG as measured with the questionnaire or with the FoG Index (King et al., under review). However, the worse the FoG index at baseline, the larger the improvement of FoG with the Agility Boot Camp, highlighting the difficulty in measuring the whole continuum of FoG. It should be noted that this study's outcomes were measured in the Off medication state and a recent study showed that the benefits of cognitive training to reduced FoG were apparent only in the On state.[96]

Highlights

- Freezing is brief, episodic inability to step with the perception that the feet are glued to the floor.
- Freezing of gait affects the majority of people with PD and is often associated with falls as the disease progresses.
- PD with FoG shows more balance disorders than nonfreezers, particularly in APAs and dynamic balance during gait.
- PD with FoG shows both structural and functional abnormalities in brain circuitry compared to nonfreezers, although it is difficult to control for severity of disease since FoG is more common as the disease progresses.
- Levodopa is the most effective treatment for FoG.
- DBS is not indicated for FoG and sometimes worsens it.
- External cueing can result in short-term improvement in reducing the duration of FoG episodes but might result in habituation.

References

1. Nutt JG, Bloem BR, Giladi N, Hallett M, Horak FB, Nieuwboer A. Freezing of gait: moving forward on a mysterious clinical phenomenon. *The Lancet Neurology* 2011;**10**(8): 734–44.
2. Youn J, Okuma Y, Hwang M, Kim D, Cho JW. Falling direction can predict the mechanism of recurrent falls in advanced Parkinson's disease. *Scientific Reports* 2017;**7**(1): 3921.
3. Paul SS, Canning CG, Sherrington C, Lord SR, Close JC, Fung VS. Three simple clinical tests to accurately predict falls in people with Parkinson's disease. *Movement Disorders: Official Journal of the Movement Disorder Society* 2013;**28**(5):655–62.
4. Hely MA, Reid WG, Adena MA, Halliday GM, Morris JG. The Sydney multicenter study of Parkinson's disease: the inevitability of dementia at 20 years. *Movement Disorders: Official Journal of the Movement Disorder Society* 2008;**23**(6):837–44.
5. Auyeung M, Tsoi TH, Mok V, et al. Ten year survival and outcomes in a prospective cohort of new onset Chinese Parkinson's disease patients. *Journal of Neurology, Neurosurgery, and Psychiatry* 2012;**83**(6):607–11.
6. Giladi N, McDermott MP, Fahn S, et al. Freezing of gait in PD: prospective assessment in the DATATOP cohort. *Neurology* 2001;**56**(12):1712–21.
7. Espay AJ, Fasano A, van Nuenen BF, Payne MM, Snijders AH, Bloem BR. "On" state freezing of gait in Parkinson disease: a paradoxical levodopa-induced complication. *Neurology* 2012;**78**(7):454–7.

8. Kerr GK, Worringham CJ, Cole MH, Lacherez PF, Wood JM, Silburn PA. Predictors of future falls in Parkinson disease. *Neurology* 2010;**75**(2):116–24.
9. Latt MD, Lord SR, Morris JG, Fung VS. Clinical and physiological assessments for elucidating falls risk in Parkinson's disease. *Movement Disorders: Official Journal of the Movement Disorder Society* 2009;**24**(9):1280–9.
10. Hausdorff JM, Balash Y, Giladi N. Time series analysis of leg movements during freezing of gait in Parkinson's disease: akinesia, rhyme or reason? *Physica A: Statistical Mechanics and Its Applications* 2003;**321**:565–70.
11. Moore ST, MacDougall HG, Ondo WG. Ambulatory monitoring of freezing of gait in Parkinson's disease. *Journal of Neuroscience Methods* 2008;**167**(2):340–8.
12. Yanagisawa N, Ueno E, Takami M. Frozen gait of Parkinson's disease and parkinsonism. A study with floor reaction forces and EMG. In: Shimamura M, Grillner S, Edgerton VR, editors. *Neurobiological basis of human locomotion*. Tokyo: Japan Scientific Societies Press; 1991.
13. Bloem BR, Hausdorff JM, Visser JE, Giladi N. Falls and freezing of gait in Parkinson's disease: a review of two interconnected, episodic phenomena. *Movement Disorders: Official Journal of the Movement Disorder Society* 2004;**19**(8):871–84.
14. Schaafsma JD, Giladi N, Balash Y, Bartels AL, Gurevich T, Hausdorff JM. Gait dynamics in Parkinson's disease: relationship to parkinsonian features, falls and response to levodopa. *Journal of the Neurological Sciences* 2003;**212**(1–2):47–53.
15. Vercruysse S, Gilat M, Shine JM, Heremans E, Lewis S, Nieuwboer A. Freezing beyond gait in Parkinson's disease: a review of current neurobehavioral evidence. *Neuroscience and Biobehavioral Reviews* 2014;**43**:213–27.
16. Vercruysse S, Spildooren J, Heremans E, et al. The neural correlates of upper limb motor blocks in Parkinson's disease and their relation to freezing of gait. *Cerebral Cortex* 2014; **24**(12):3154–66.
17. Cohen RG, Klein KA, Nomura M, et al. Inhibition, executive function, and freezing of gait. *Journal of Parkinson's Disease* 2014;**4**(1):111–22.
18. Bekkers EMJ, Dijkstra BW, Dockx K, Heremans E, Verschueren SMP, Nieuwboer A. Clinical balance scales indicate worse postural control in people with Parkinson's disease who exhibit freezing of gait compared to those who do not: a meta-analysis. *Gait and Posture* 2017;**56**:134–40.
19. Schlenstedt C, Mancini M, Nutt J, et al. Are hypometric anticipatory postural adjustments contributing to freezing of gait in Parkinson's disease? *Frontiers in Aging Neuroscience* 2018;**10**:36.
20. Jacobs JV, Nutt JG, Carlson-Kuhta P, Stephens M, Horak FB. Knee trembling during freezing of gait represents multiple anticipatory postural adjustments. Experimental Neurology 2009; 215(2): 334-341.
21. Smulders K, Esselink RA, De Swart BJ, Geurts AC, Bloem BR, Weerdesteyn V. Postural inflexibility in PD: does it affect compensatory stepping? *Gait and Posture* 2014;**39**(2): 700–6.
22. Peterson DS, Horak FB. Effects of freezing of gait on postural motor learning in people with Parkinson's disease. *Neuroscience* 2016;**334**:283–9.
23. El-Gohary M, Peterson D, Gera G, Horak FB, Huisinga JM. Validity of the instrumented Push and release test to quantify postural responses in persons with multiple sclerosis. *Archives of Physical Medicine and Rehabilitation* 2017;**98**(7):1325–31.
24. Smith BA, Carlson-Kuhta P, Horak FB. Consistency in administration and response for the backward Push and release test: a clinical assessment of postural responses. *Physiotherapy Research International: The Journal for Researchers and Clinicians in Physical Therapy* 2016;**21**(1):36–46.
25. Jacobs JV, Horak FB, Van Tran K, Nutt JG. An alternative clinical postural stability test for patients with Parkinson's disease. *Journal of Neurology* 2006;**253**(11):1404–13.

26. Schieppati M, Nardone A. Free and supported stance in Parkinson's disease. The effect of posture and 'postural set' on leg muscle responses to perturbation, and its relation to the severity of the disease. *Brain: A Journal of Neurology* 1991;**114**(Pt 3):1227–44.
27. Nantel J, Bronte-Stewart H. The effect of medication and the role of postural instability in different components of freezing of gait (FOG). *Parkinsonism and Related Disorders* 2014;**20**(4):447–51.
28. Pelykh O, Klein AM, Botzel K, Kosutzka Z, Ilmberger J. Dynamics of postural control in Parkinson patients with and without symptoms of freezing of gait. *Gait and Posture* 2015;**42**(3):246–50.
29. Vervoort G, Bengevoord A, Strouwen C, et al. Progression of postural control and gait deficits in Parkinson's disease and freezing of gait: a longitudinal study. *Parkinsonism and Related Disorders* 2016;**28**:73–9.
30. Huh YE, Hwang S, Kim K, Chung WH, Youn J, Cho JW. Postural sensory correlates of freezing of gait in Parkinson's disease. *Parkinsonism and Related Disorders* 2016;**25**:72–7.
31. Schlenstedt C, Muthuraman M, Witt K, Weisser B, Fasano A, Deuschl G. Postural control and freezing of gait in Parkinson's disease. *Parkinsonism and Related Disorders* 2016;**24**:107–12.
32. de Souza Fortaleza AC, Mancini M, Carlson-Kuhta P, et al. Dual task interference on postural sway, postural transitions and gait in people with Parkinson's disease and freezing of gait. *Gait and Posture* 2017;**56**:76–81.
33. Peterka RJ. Sensorimotor integration in human postural control. *Journal of Neurophysiology* 2002;**88**(3):1097–118.
33a. Feller KJ, Peterka RJ, Horak FB. Sensory re-weighting for postural control in Parkinson's disease. *Front Hum Neurosci* 2019;**13**:126. https://doi.org/10.3389/fnhum.2019.00126. eCollection 2019.
34. Maurer C. Postural deficits in Parkinson's disease are caused by insufficient feedback motor error correction and by deficits in sensory reweighting. *Clinical Neurophysiology* 2009;**120**:e82.
35. Plotnik M, Giladi N, Balash Y, Peretz C, Hausdorff JM. Is freezing of gait in Parkinson's disease related to asymmetric motor function? *Annals of Neurology* 2005;**57**(5):656–63.
36. Plotnik M, Giladi N, Hausdorff JM. Bilateral coordination of walking and freezing of gait in Parkinson's disease. *European Journal of Neuroscience* 2008;**27**(8):1999–2006.
37. Plotnik M, Hausdorff JM. The role of gait rhythmicity and bilateral coordination of stepping in the pathophysiology of freezing of gait in Parkinson's disease. *Movement Disorders: Official Journal of the Movement Disorder Society* 2008;**23**(Suppl. 2):S444–50.
38. Hausdorff JM, Schaafsma JD, Balash Y, Bartels AL, Gurevich T, Giladi N. Impaired regulation of stride variability in Parkinson's disease subjects with freezing of gait. *Experimental Brain Research* 2003;**149**(2):187–94.
39. Weiss A, Herman T, Giladi N, Hausdorff JM. New evidence for gait abnormalities among Parkinson's disease patients who suffer from freezing of gait: insights using a body-fixed sensor worn for 3 days. *Journal of Neural Transmission* 2015;**122**(3):403–10.
40. Iansek R, Huxham F, McGinley J. The sequence effect and gait festination in Parkinson disease: contributors to freezing of gait? *Movement Disorders: Official Journal of the Movement Disorder Society* 2006;**21**(9):1419–24.
41. Peterson DS, Fling BW, Mancini M, Cohen RG, Nutt JG, Horak FB. Dual-task interference and brain structural connectivity in people with Parkinson's disease who freeze. *Journal of Neurology, Neurosurgery, and Psychiatry* 2015;**86**(7):786–92.
42. Shine JM, Matar E, Ward PB, et al. Freezing of gait in Parkinson's disease is associated with functional decoupling between the cognitive control network and the basal ganglia. *Brain: A Journal of Neurology* 2013;**136**(Pt 12):3671–81.
43. Maidan I, Bernad-Elazari H, Gazit E, Giladi N, Hausdorff JM, Mirelman A. Changes in oxygenated hemoglobin link freezing of gait to frontal activation in patients with

Parkinson disease: an fNIRS study of transient motor-cognitive failures. *Journal of Neurology* 2015;**262**(4):899–908.
43a. Mancini M, Bloem BR, Horak FB, Lewis SJG, Nieuwboer A, Nonnekes J. Clinical and methodological challenges for assessing freezing of gait: Future perspectives. *Mov Disord* 2019;**34**(6):783–90. https://doi.org/10.1002/mds.27709. Epub 2019 May 2.
44. Mancini M, Smulders K, Cohen RG, Horak FB, Giladi N, Nutt JG. The clinical significance of freezing while turning in Parkinson's disease. *Neuroscience* 2017;**343**:222–8.
45. Nieuwboer A, Rochester L, Herman T, et al. Reliability of the new freezing of gait questionnaire: agreement between patients with Parkinson's disease and their carers. *Gait and Posture* 2009;**30**(4):459–63.
46. Fasano A, Herman T, Tessitore A, Strafella AP, Bohnen NI. Neuroimaging of freezing of gait. *Journal of Parkinson's Disease* 2015;**5**(2):241–54.
47. Fasano A, Laganiere SE, Lam S, Fox MD. Lesions causing freezing of gait localize to a cerebellar functional network. *Annals of Neurology* 2017;**81**(1):129–41.
48. Fling BW, Cohen RG, Mancini M, et al. Functional reorganization of the locomotor network in Parkinson patients with freezing of gait. *PLoS One* 2014;**9**(6):e100291.
49. Wang M, Jiang S, Yuan Y, et al. Alterations of functional and structural connectivity of freezing of gait in Parkinson's disease. *Journal of Neurology* 2016;**263**(8):1583–92.
50. Canu E, Agosta F, Sarasso E, et al. Brain structural and functional connectivity in Parkinson's disease with freezing of gait. *Human Brain Mapping* 2015;**36**(12):5064–78.
51. Tessitore A, Amboni M, Esposito F, et al. Resting-state brain connectivity in patients with Parkinson's disease and freezing of gait. *Parkinsonism and Related Disorders* 2012;**18**(6): 781–7.
52. Fling BW, Cohen RG, Mancini M, Nutt JG, Fair DA, Horak FB. Asymmetric pedunculopontine network connectivity in parkinsonian patients with freezing of gait. *Brain: A Journal of Neurology* 2013;**136**(Pt 8):2405–18.
53. Cohen RG, Horak FB, Nutt JG. Peering through the FoG: visual manipulations shed light on freezing of gait. *Movement Disorders* 2012;**27**(4):470–2.
54. Shine JM, Matar E, Ward PB, et al. Exploring the cortical and subcortical functional magnetic resonance imaging changes associated with freezing in Parkinson's disease. *Brain: A Journal of Neurology* 2013;**136**(Pt 4):1204–15.
55. Lewis SJ, Barker RA. A pathophysiological model of freezing of gait in Parkinson's disease. *Parkinsonism and Related Disorders* 2009;**15**(5):333–8.
56. Lewis SJ, Shine JM. The next step: a common neural mechanism for freezing of gait. *The Neuroscientist: A Review Journal Bringing Neurobiology, Neurology and Psychiatry* 2016; **22**(1):72–82.
57. Gilat M, Ehgoetz Martens KA, Miranda-Dominguez O, et al. Dysfunctional limbic circuitry underlying freezing of gait in Parkinson's disease. *Neuroscience* 2018;**374**:119–32.
58. Burn DJ, Landau S, Hindle JV, et al. Parkinson's disease motor subtypes and mood. *Movement Disorders: Official Journal of the Movement Disorder Society* 2012;**27**(3):379–86.
59. Ehgoetz Martens KA, Ellard CG, Almeida QJ. Does anxiety cause freezing of gait in Parkinson's disease? *PLoS One* 2014;**9**(9):e106561.
60. Giladi N, Hausdorff JM. The role of mental function in the pathogenesis of freezing of gait in Parkinson's disease. *Journal of the Neurological Sciences* 2006;**248**(1–2):173–6.
61. Lieberman A. Are freezing of gait (FOG) and panic related? *Journal of the Neurological Sciences* 2006;**248**(1–2):219–22.
62. Ehgoetz Martens KA, Pieruccini-Faria F, Almeida QJ. Could sensory mechanisms be a core factor that underlies freezing of gait in Parkinson's disease? *PLoS One* 2013;**8**(5): e62602.
63. Maidan I, Plotnik M, Mirelman A, Weiss A, Giladi N, Hausdorff JM. Heart rate changes during freezing of gait in patients with Parkinson's disease. *Movement Disorders: Official Journal of the Movement Disorder Society* 2010;**25**(14):2346–54.

64. Mazilu S, Calatroni A, Gazit E, Mirelman A, Hausdorff JM, Troster G. Prediction of freezing of gait in Parkinson's from physiological wearables: an exploratory study. *IEEE Journal of Biomedical and Health Informatics* 2015;**19**(6):1843–54.
65. Ehgoetz Martens KA, Lukasik EL, Georgiades MJ, et al. Predicting the onset of freezing of gait: a longitudinal study. *Movement Disorders: Official Journal of the Movement Disorder Society* 2018;**33**(1):128–35.
66. De Witte NAJ, Mueller SC. White matter integrity in brain networks relevant to anxiety and depression: evidence from the human connectome project dataset. *Brain Imaging and Behavior* 2017;**11**(6):1604–15.
67. Marchand WR. Cortico-basal ganglia circuitry: a review of key research and implications for functional connectivity studies of mood and anxiety disorders. *Brain Structure and Function* 2010;**215**(2):73–96.
68. Ochsner KN, Silvers JA, Buhle JT. Functional imaging studies of emotion regulation: a synthetic review and evolving model of the cognitive control of emotion. *Annals of the New York Academy of Sciences* 2012;**1251**:E1–24.
69. Rohr CS, Dreyer FR, Aderka IM, et al. Individual differences in common factors of emotional traits and executive functions predict functional connectivity of the amygdala. *NeuroImage* 2015;**120**:154–63.
70. Amboni M, Cozzolino A, Longo K, Picillo M, Barone P. Freezing of gait and executive functions in patients with Parkinson's disease. *Movement Disorders: Official Journal of the Movement Disorder Society* 2008;**23**(3):395–400.
71. Brugger F, Abela E, Hagele-Link S, Bohlhalter S, Galovic M, Kagi G. Do executive dysfunction and freezing of gait in Parkinson's disease share the same neuroanatomical correlates? *Journal of the Neurological Sciences* 2015;**356**(1–2):184–7.
72. Walton CC, O'Callaghan C, Hall JM, et al. Antisaccade errors reveal cognitive control deficits in Parkinson's disease with freezing of gait. *Journal of Neurology* 2015;**262**(12): 2745–54.
73. Vandenbossche J, Deroost N, Soetens E, et al. Freezing of gait in Parkinson's disease: disturbances in automaticity and control. *Frontiers in Human Neuroscience* 2012;**6**:356.
74. Lenka A, Naduthota RM, Jha M, et al. Freezing of gait in Parkinson's disease is associated with altered functional brain connectivity. *Parkinsonism and Related Disorders* 2016; **24**:100–6.
75. Delval A, Moreau C, Bleuse S, et al. Gait and attentional performance in freezers under methylphenidate. *Gait and Posture* 2015;**41**(2):384–8.
76. Espay AJ, Dwivedi AK, Payne M, et al. Methylphenidate for gait impairment in Parkinson disease: a randomized clinical trial. *Neurology* 2011;**76**(14):1256–62.
77. Gurevich T, Peretz C, Moore O, Weizmann N, Giladi N. The effect of injecting botulinum toxin type a into the calf muscles on freezing of gait in Parkinson's disease: a double blind placebo-controlled pilot study. *Movement Disorders: Official Journal of the Movement Disorder Society* 2007;**22**(6):880–3.
78. Wieler M, Camicioli R, Jones CA, Martin WR. Botulinum toxin injections do not improve freezing of gait in Parkinson disease. *Neurology* 2005;**65**(4):626–8.
79. Giladi N, Gurevich T, Shabtai H, Paleacu D, Simon ES. The effect of botulinum toxin injections to the calf muscles on freezing of gait in parkinsonism: a pilot study. *Journal of Neurology* 2001;**248**(7):572–6.
80. Karachi C, Cormier-Dequaire F, Grabli D, et al. Clinical and anatomical predictors for freezing of gait and falls after subthalamic deep brain stimulation in Parkinson's disease patients. *Parkinsonism and Related Disorders* 2019.
81. Schrader C, Capelle HH, Kinfe TM, et al. GPi-DBS may induce a hypokinetic gait disorder with freezing of gait in patients with dystonia. *Neurology* 2011;**77**(5):483–8.

82. Huang C, Chu H, Zhang Y, Wang X. Deep brain stimulation to alleviate freezing of gait and cognitive dysfunction in Parkinson's disease: update on current research and future perspectives. *Frontiers in Neuroscience* 2018;**12**:29.
83. Su D, Chen H, Hu W, et al. Frequency-dependent effects of subthalamic deep brain stimulation on motor symptoms in Parkinson's disease: a meta-analysis of controlled trials. *Scientific Reports* 2018;**8**(1):14456.
84. Ginis P, Nackaerts E, Nieuwboer A, Heremans E. Cueing for people with Parkinson's disease with freezing of gait: a narrative review of the state-of-the-art and novel perspectives. *Annals of Physical and Rehabilitation Medicine* 2018;**61**(6):407–13.
85. Barthel C, Nonnekes J, van Helvert M, et al. The laser shoes: a new ambulatory device to alleviate freezing of gait in Parkinson disease. *Neurology* 2018;**90**(2):e164–71.
86. Barthel C, van Helvert M, Haan R, et al. Visual cueing using laser shoes reduces freezing of gait in Parkinson's patients at home. *Movement Disorders: Official Journal of the Movement Disorder Society* 2018;**33**(10):1664–5.
87. Casamassima F, Ferrari A, Milosevic B, Ginis P, Farella E, Rocchi L. A wearable system for gait training in subjects with Parkinson's disease. *Sensors* 2014;**14**(4):6229–46.
88. Harrington W, Greenberg A, King E, et al. Alleviating freezing of gait using phase-dependent tactile biofeedback. In: *Conference proceedings: annual international conference of the IEEE engineering in medicine and biology society IEEE engineering in medicine and biology society annual conference*, vol. 2016; 2016. p. 5841–4.
89. Mancini M, Smulders K, Harker G, Stuart S, Nutt JG. Assessment of the ability of open- and closed-loop cueing to improve turning and freezing in people with Parkinson's disease. *Scientific Reports* 2018;**8**(1):12773.
90. Stummer C, Dibilio V, Overeem S, Weerdesteyn V, Bloem BR, Nonnekes J. The walk-bicycle: a new assistive device for Parkinson's patients with freezing of gait? *Parkinsonism and Related Disorders* 2015;**21**(7):755–7.
91. Killane I, Fearon C, Newman L, et al. Dual motor-cognitive virtual reality training impacts dual-task performance in freezing of gait. *IEEE Journal of Biomedical Health Informatics* 2015;**19**(6):1855–61.
92. Hackney ME, Earhart GM. Effects of dance on movement control in Parkinson's disease: a comparison of Argentine tango and American ballroom. *Journal of Rehabilitation Medicine* 2009;**41**(6):475–81.
93. Duncan RP, Earhart GM. Randomized controlled trial of community-based dancing to modify disease progression in Parkinson disease. *Neurorehabilitation and Neural Repair* 2012;**26**(2):132–43.
94. dos Santos Delabary M, Komeroski IG, Monteiro EP, Costa RR, Haas AN. Effects of dance practice on functional mobility, motor symptoms and quality of life in people with Parkinson's disease: a systematic review with meta-analysis. *Aging Clinical and Experimental Research* 2018;**30**(7):727–35.
95. Mirelman A, Maidan I, Herman T, Deutsch JE, Giladi N, Hausdorff JM. Virtual reality for gait training: can it induce motor learning to enhance complex walking and reduce fall risk in patients with Parkinson's disease? *The Journals of Gerontology Series A, Biological Sciences and Medical Sciences* 2011;**66**(2):234–40.
96. Walton CC, Shine JM, Mowszowski L, Naismith SL, Lewis SJ. Freezing of gait in Parkinson's disease: current treatments and the potential role for cognitive training. *Restorative Neurology and Neuroscience* 2014;**32**(3):411–22.

CHAPTER

9

How should the clinician approach imbalance in PD?

A. When in the course of PD should clinicians address balance issues?

Clinicians should address balance issues at every stage (milestone) of the disease. Treatment of balance disorders in Parkinson disease (PD) can be thought of in terms of milestones of the disease emphasizing clinical aspects of imbalance that become more and more prominent as the disease progresses and the interventions that may be considered. Milestones of PD balance deficits and potential treatments are indicated in Table 9.1. We will consider the treatment of balance at these six stages realizing that these stages and their therapy may overlap. Care of balance disorders during the course of the disease is most effective through the coordination of medical and physical therapy care providers. We will discuss the considerations for both medical and physical therapy care providers at each stage. The focus of this review is on balance and not on other motor or nonmotor features of PD.

B. When is exercise helpful for balance disorders?

Milestone 1. Early PD with no evident balance or gait deficits

At this stage of the disease, the person with PD may have minor parkinsonian signs, perhaps just mild tremor, bradykinesia, and rigidity in one arm and no clinically apparent problems with balance. However, more sophisticated studies find that even at this stage there are differences in early PD and age-matched healthy controls. People with PD are less active at home at this early stage: they take fewer steps per day[1] and

Balance Dysfunction in Parkinson's Disease
https://doi.org/10.1016/B978-0-12-813874-8.00009-X

TABLE 9.1 Milestones of Parkinson disease (PD) balance disorders and treatments.

1	**Early PD: No symptoms or overt evidence of balance deficits**	**Aerobic, strength, and balance exercise. Building habit of exercise in a community**
2	Slowing of gait, sense of imbalance, near falls: levodopa effects	Exercise targeted to the specific balance impairment and fall prevention
3	Falls	Continuing exercise and home assessment
		Falls and injury reduction
4	Increasing motor deficits: Consideration of deep brain stimulation (DBS)	Strategies to prevent falls in Off and On state Site of DBS matters
		Exercising and cueing to overcome FoG
		Cognitive training
5	Freezing of gait (FoG)	Anxiety reduction
		Alter home risk environment
6	Unable to ambulate without assistance	Practicing transfers

their movements during the day are of shorter duration than age-matched controls subjects.[2] These measures suggest that even in early PD, people with the disease are less fit. This observation of reduced physical activity in early PD may correlate with the epidemiologic studies indicating an increased risk of PD in people that are physically inactive in midlife.[3,4] Although it is possible that reduced physical activity is a consequence of preclinical PD,[5] this explanation has been discounted in recent studies examining the lag between active exercise and onset of PD.[6] Thus, there is great interest in physical activity throughout adult life which may delay or reduce the risk of PD.[5]

Many people with early PD perform the Mini-BESTest, Berg Balance test, and the Timed Up and Go (TUG) less well than controls and may have a sense of imbalance.[7] Early, untreated PD patients have increased postural sway when standing (see Chapter 3) indicating impaired standing balance which is associated with increased chance of falling. Dynamic balance measured as increased gait variability, another measure of increased risk for falling, is also impaired even in early, untreated PD,[8] Chapter 6. Turns while walking are slower than normal which may represent an automatic compensation for impaired dynamic balance (Chapter 7). In accord with these studies, falls are more frequent in people with early PD than in otherwise healthy elderly people.[9,10]

The neurological care provider encountering a patient at this stage should confirm the diagnosis of PD by history and exam. If uncertainty exists, further testing with dopamine transporter (DaT) scan, brain MRI, or other studies could be considered. Other comorbidities that might contribute to imbalance such as impaired vision, vertigo, arthritis, and peripheral neuropathy should be identified and appropriate diagnostics, referrals, and treatments instituted. Education of the patient about PD is essential and booklets, web sites, and classes directed at the newly diagnosed patient should be recommended.

A critical aspect of this initial, clinical contact is to emphasize the importance of activity and exercise in delaying balance and mobility issues. There is rationale and some evidence that physical activity may delay the progression of PD, that is, exercise may be a neuroprotective intervention.[5] A 2-year randomized pilot study with intensive rehabilitation treatment found that the 16 subjects in the active arm had slower progression of the UPDRS II and III, Timed Up and Go (TUG), and Parkinson's Disease Disability Scale (PPDS) and the levodopa equivalents were about half those in the 15 control subjects.[11] Further, a randomized clinical trial in early, untreated PD patients found that high-intensity exercise reduced the increase in the total UPDRS score over 6 months compared to the low-intensity exercise group.[12] In patients that are already treated, progressive resistance exercise reduced progression of the motor UPDRS scores over 24 months compared to a general, nonprogressive exercise program.[13]

Because early PD subjects are already at risk for falls or may have fallen,[9,10] exercise to reduce or prevent falling is indicated. Exercise programs containing strengthening and aerobic exercises will increase strength and aerobic fitness in PD but have minimal effects on balance and gait.[14] On the other hand, exercise programs that emphasize balance and gait do reduce falls for up to 12 months (Fig. 9.1).[15–17] The effects of exercise are robust for facility-based and supervised "balance-centric" exercise with less or no effects for home-based exercise programs.[18] Likewise, activities that include challenges to balance are effective. Dance improved Motor UPDRS and TUG[19] and Tai chi also improved balance measures and reduced falls relative to a control stretching group (Fig. 9.1).[20,21]

In summary, there is suggestive evidence that exercise may reduce progression of PD and strong evidence that supervised challenging balance exercises reduce falling. For these reasons, patients should be referred to physical therapist (PT) for balance-specific exercises and to PD exercise classes emphasizing balance. If possible, the patient should participate in some community exercise programs as well as exercise on their own. The patient should be provided with concrete guidelines for exercise (see Table 9.2). The Physical Activity Guidelines for Americans

Effects on fall rate in long-term (n=451)

Study name (follow-up period)	Outcome	Rate ratio	Lower limit	Upper limit	p-Value
Gao 2014 (26 wk)	fall rate	0.469	0.231	0.951	0.036
Goodwin 2011 (10 wk)	fall rate	0.740	0.411	1.333	0.316
Li 2012 (balance ex, 12 wk)	fall rate	0.310	0.142	0.678	0.003
Li 2012 (strength ex, 12 wk)	fall rate	0.400	0.181	0.884	0.024
Morris 2015 (strength ex, 12 m)	fall rate	0.151	0.071	0.322	0.000
Smania 2010 (4 wk)	fall rate	0.559	0.371	0.842	0.005
		0.413	0.270	0.630	0.000

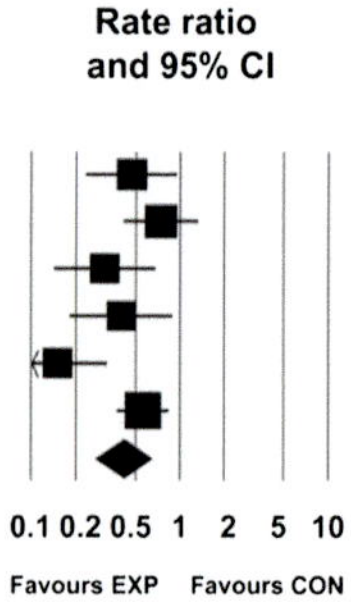

Statistics for each study

Rate ratio and 95% CI

0.1 0.2 0.5 1 2 5 10

Favours EXP Favours CON

Random-effects model of meta-analysis: $I^2=61\%$, df=5, $P<0.05$

FIGURE 9.1 Forest plots from meta-analyses in Shen et al.[16] of the effects of exercise training on fall rate in people with PD demonstrating estimates of effect size with 95% confidence intervals (CIs).

TABLE 9.2 Few reliable sources of information for exercise for PD.

Foundation	Website
Parkinson's Foundation	https://parkinson.org/
The Michael J Fox Foundation	https://www.michaelifox.org/
Parkinson UK	https://www.parkinsons.org.uk/
Davis Phinney Foundation	https://www.davisphinneyfoundation.org/
Brian Grant Foundation	https://briangrant.org/

may be useful for clinicians and patients. Emphasize to the patient and to the spouse/partner that they must make exercise a habit.

In early PD, the PT care provider should assess the fitness, balance and gait, and the influence of any comorbidities. Based on these assessments, the PT provider should develop a home exercise program and help them identify appropriate, evidence-based community exercise programs led by exercise trainers who are knowledgeable about working with people who have PD. The program should be specific, tailored to the patient's preferences and activities that the patient already does. Also, the PT and neurological care provider should emphasize the reliable resources to learn about exercise and warn about unproven gimmicks, foods and other products advertised in various media.

C. Does levodopa, dopamine replacement therapy, improve balance?

Milestone 2. Slowing of gait, sense of imbalance, and near falls—initiating levodopa

As the disease progresses, patients begin to describe slowing of gait, a sense of imbalance, and near falls. The neurological care provider should observe the gait in a hallway for slowness. Particular attention should be paid to turns because slowing, shortening of steps, and extra steps to maintain balance during turning may be evident when the straight-ahead walking appears normal. Other postural transitions such as arising from a chair or bed and rolling should also be observed. Clinicians should pay attention to how the patient moves when they are not aware they are being observed. The patient's walking and postural transitions may be much better when they are directing their attention to their movement than when their attention is diverted and they are not aware they are being watched. Balance should be also assessed via the response to the Push and Release test (Chapter 4). In the Push and Release test, the standing patient leans back against the examiner's hands until the center of mass is behind the heels. The examiner then quickly pulls their hands away and the person must take a rapid step to regain their balance. For the patient's and the examiner's safety, there should be a wall behind the examiner to prevent falling. The Push and Release test is more sensitive to imbalance (it may be abnormal when the Pull test is not) and can be more uniformly performed.[22] The Pull test of the standard UPDRS exam is influenced by the suddenness and force of the pull, as well as the patient may anticipate the pull and lean forward to counter the pull. The push and release test identifies patients that will fall better than does the Pull test.[23]

Abnormalities in balance and walking may be reasons to start levodopa if other parkinsonian motor signs, such as limb bradykinesia and rigidity, have not already been a reason to start therapy. If levodopa has already been initiated, then altering the dosing might be appropriate. Although there are other antiparkinsonian agents, when balance and gait are affected, levodopa is the drug of choice because of its greater efficacy.

What are the balance features that may respond to levodopa? They vary and relate to severity of disease and what aspects of balance are examined. However, both bradykinesia and rigidity impair most domains of balance and are responsive to levodopa. Although tremor can contribute to increased postural sway, the response to levodopa is more erratic and many patients get no benefit in tremor when bradykinesia and rigidity do respond. Chapter 3 describes how functional limits of stability are increased by levodopa. Standing postural sway is increased in PD, even early in the disease even before treatment. This increased standing sway during stance may initially be reduced by levodopa in early PD.[24] However, with continued levodopa therapy and progression of the disease, sway will increase with levodopa. The increased sway may be related to dyskinesia[25] which often starts with slight rocking of the head or trunk that patient and caregivers may not be aware of. Standing sway has even been used as a measure of levodopa-induced dyskinesia.[26] Dyskinesia is associated with falls[27] presumably via the increased sway which is associated with increased fall risk.[25] In addition, dynamic stability, described in Chapter 6, measured with gait variability is generally reduced by levodopa, although it does not normalize.[28]

Reactive postural responses in PD are hypoactive (bradykinetic), as described in Chapter 4. Onset of postural responses is not slow; latency to muscle response to a perturbation is unchanged. However, the muscle activation observed by EMG patterns is less organized and the resultant torque is slower to develop and is reduced compared to normal individuals. Levodopa may increase the force generated by the postural responses but does not return them to normal. Switching between postural response strategies when the environmental conditions change is difficult for people with PD and does not improve with levodopa. Anticipatory postural adjustments prior to step initiation, described in Chapter 5, are augmented with levodopa and increased gait speed and stride length are the most obvious gait clinical improvements with levodopa described in Chapter 6. Thus the effects of levodopa on balance depend upon what aspects of balance are examined.

At this stage, the physical therapist assesses the patient's anticipatory and reactive postural responses, sensory orientation, and challenged gait with the Mini-BESTest.[29] For example, if postural responses are impaired, the PT can work on compensatory stepping to produce large and quick postural adjustments. For impaired sensory orientation, the therapist can have the patient practice in situations with reduced sensory input from support surface or visual surroundings. The patient can be encouraged to walk on uneven terrain such as the yard, garden, or trails. Practicing balance and walking tasks with concurrent activities helps to keep balance control automatic and prepares the patient for the real world where most balance and walking are carried out while doing other tasks or talking and listening to other people. Finally, the PT should review and revise the exercise program to address the patient's balance deficits at this stage of the disease.

D. What considerations should a history of falls trigger?

Milestone 3. Falls

Falls put the patient at risk for injury but also affect the person's confidence in their balance and, in turn, their mobility and participation in home and community activities. It is important to search for clues to the etiology of the falls by eliciting a detailed history of the fall. Table 9.3 summarizes 3Cs for evaluating fall history in order to prevent future falls: Circumstances, Characterization, and Consequences.[30] For example, exactly where was the patient? in the home or community? What clothes and particularly footwear were they wearing? What were they doing immediately preceding the fall? Were there was any symptoms or premonition that they were about to fall (dizziness or light headedness)? An important distinction is whether the patient collapsed suggesting loss of axial tone or toppled like a falling tree, implying retained postural tone? Collapsing with loss of tone suggests loss of consciousness that can occur with orthostatic hypotension or cardiac arrhythmias. Was there loss of consciousness and how did they get up from the surface to which they fell? Another consideration is response to antiparkinsonian medications at the time of the fall. Was the patient slow or OFF, suggesting that the patient might have been underdosed or having motor fluctuations in

TABLE 9.3 The three Cs for falls in Parkinson disease.

- ✓ Circumstances
  - Walking, standing, transitions
  - Parkinsonian signs: ON or OFF dyskinesia
  - Use of walking devices
  - Distraction: carrying something, talking, urinary urgency, other
  - Location: Where the fall takes place, stairs
  - Lighting
  - Preceding symptoms: vertigo, lightheadedness
  - Previous falls?
- ✓ Characterization
  - Trip, slip, loss of balance, freezing
  - Direction of fall. Forward (walking), Backward (standing or backing up), Side (turning)
  - Collapsing (loss of postural tone)
  - Loss of consciousness
  - How did subject arise from fall
- ✓ Consequences
  - Injuries
  - Activity restriction (fear of falling)

Adapted from Fasano A, Canning CG, Hausdorff JM, Lord S, Rochester L. Falls in Parkinson's disease: a complex and evolving picture. Movement Disorders: Official Journal of the Movement Disorder Society *2017;***32***(11): 1524–36.*

response to infrequent doses of medication? If, when ON with levodopa therapy, balance and gait are sufficient for safe mobility, strategies to produce more continuous levodopa effects may be considered, such as adjusting the doses or frequency of levodopa doses, enteral infusion of levodopa, use of dopamine extenders or rescue therapy. On the other hand, did the patient have levodopa-induced dyskinesia at the time of falls since dyskinesia increases sway and the risk of falls?[27] If so, altering levodopa dosing or adding an antidyskinetic agent such as amantadine may reduce the fall risk.

A strong, interdisciplinary team is needed to address the many, treatable risk factors for falls. Table 9.4 summarizes many of the treatable risk factors for falls in people with PD, as well as treatable comorbidities common in the elderly (such as vestibular disorders) that can be addressed by the team. One of the most predictive factors for future falls is a history of falls, but falls can be reduced by reducing the total number of treatable risk factors (Tinetti). A comprehensive screening for treatable risk factors by neurologists, internists, nurses, psychologists, therapists, and others can reduce future fall risk and thereby improve quality of life. Reducing falls also increases the lifespan since consequences of falls is the most common cause of mortality in older people, especially older people with PD.

A cross-sectional analysis of possible causes of falls in 86 people with PD[27] found that fallers had longer disease duration, higher Levodopa-equivalent doses, greater On time with dyskinesia (all $P < .005$), and higher scores on some Movement Disorder Society-Unified Parkinson's Disease Rating Scale items, particularly axial scores. Surpisingly, patients with falls did not differ from nonfallers in age or overall motor UPDRS scores. Severity of psychosis, executive cognitive impairment, autonomic (particularly cardiovascular) dysfunction, and sleep disturbances (particularly REM sleep behavioral disorder) were also significantly associated with falls. Medications need to be screened because fallers more frequently reported use of antidepressants (both tricyclics and SSRIs) and neuroleptics, but not hypnotics. In logistic regression analysis, cardiovascular dysfunction, antidepressant use, and REM sleep behavioral disorder were also significantly associated with falls.

The physical exam neurologists should search for changes in parkinsonism (signs suggestive of a parkinsonism plus disorder or dystonia of foot or leg), other neurological disorders (stroke, subdural hematoma, peripheral nerve and root disorders, or vestibular dysfunction) or emergence of other systemic disorders (cardiac and musculoskeletal diseases). As orthostatic hypotension is common in PD and treatable, blood pressures sitting and standing should be checked. Although lying and standing blood pressures may be the best method to detect orthostatic hypotension, this is rarely possible in a busy clinic. It is important to

TABLE 9.4 Treatable risk factors for falls.

Risk factor	Intervention
Environmental lighting, clutter, rugs	•Home visits •Photos
Assistive Devices canes, walkers	•PT to evaluate proper use
Musculoskeletal Disorders back, hip, knee	•Evaluation •Physical therapy
Anxiety and Depression	•Counseling •Non cholinergic antidepressant
Orthostatic hypotension	•Evaluate drug etiology •Treat
Vision glasses, bifocals	•Ophthalmologist to evaluate and treat
Hearing and Vertigo	•Evaluation • Vestibular rehabilitation
Medications Benzodiazepines, anticholinergics	•Review and eliminate
Cognition judgment, distractibility	•Cognitive Rehabilitation •Cholinesterase inhibitors
Pain	•Massage, education, exercise
Urinary Incontinence	• Medication, bedside urinal

recognize that one test of sitting and standing blood pressures may not detect orthostatic hypotension even if it is a cause of falls. Orthostatic hypotension may vary during the day and be related to meals, levodopa or dopamine agonist intake, and to other medications in the patient's drug regimen. Having the caregiver record sitting and standing blood pressures at home when the patient notices lightheadedness or times when orthostatic hypotension is predicted to be more likely, such as after antiparkinsonian medication intake and after meals, may capture the phenomenon.

At this stage, the care provider should also assess the cognition, attention, insight, impulsivity, and mood as all of these features can affect balance and walking. Cognitive rehabilitation[31] may assist with cognitive changes, attention, and impulsivity (acting before thinking). Physical exercise also improves global cognitive function and executive function.[32]

Carefully reviewing the patient's medications that can potentially induce confusion, sedation, memory deficits, and hypotension is also important. Anticholinergic drugs are always suspect. Anticholinergics include peripherally active anticholinergics for bladder frequency and urgency, antidepressants with anticholinergic properties, first generation antihistaminics such as Benadryl, amantadine, and other anticholinergics used to treat PD. Benzodiazepines, can affect cognition and attention and are associated with falls in the elderly.[33] Antidepressants have also been associated with falls in the elderly.[34] Antihypertensive medications and medications used for urinary hesitancy, alpha-adrenergic blockers such as terazosin and doxazosin, can contribute to orthostatic hypotension and falls. Impulsivity, often with obsessive behavior, can be related to dopamine agonists. Finally, alcohol use should always be considered as a risk factor for falls.

The physical therapist will also take a history of falls. The patient will be quizzed about environment in which they fall, clothing and footwear, activity preceding the fall, what direction they fall, and how they arise after a fall. A home evaluation may allow for recommendations to improve the home safety for mobility. For example, if falls often occur at night when the patient arises in the OFF medication state to walk to the bathroom, a bedside urinal and medication to reduce urgency can be helpful. If home evaluations are not feasible, photos of the home interior and particularly places where the patient falls or has difficulty moving about may be helpful. This information, combined with comprehensive assessment of balance, will direct treatment. Examining the patient in clinic when he or she has come to a morning appointment without taking his or her morning dose of antiparkinsonian medications and is Off and again later when antiparkinsonian medications are working (On) will determine the ability of levodopa to affect balance and mobility.

Treating pain, weakness, and poor fitness are important for any elderly person with a history of falls. In addition, balance training focuses on the particular domains of balance control affected in each patient. For example, practicing compensatory stepping, turning safely, arising from a chair, as well as sitting down and arising from the floor, are often appropriate exercises. In the past decade, many studies have shown that automatic postural responses in response to unexpected perturbations can be improved with practice in the elderly and in people with PD.[35] Also, teaching strategies for dealing with multiple concurrent activities and safely turning while walking can improve automaticity of mobility.

Hiking poles may be considered to encourage walking more quickly and on more challenging surfaces. The home or class exercise programs should be adapted to changes in balance. Finally, it is important to reemphasize the importance of remaining active and continuing to exercise. Involving the caregiver in these aspects of therapy is helpful as the caregiver can remind the patient of the principles of safe mobility. At this point, the therapist should consider assistive devices for the home and in the community. A light-weight, folding wheelchair may allow the patient to remain mobile in public spaces.

E. How will deep brain stimulation affect balance?

Milestone 4. Considering Deep Brain Stimulation as treatment strategy for balance and gait dysfunction

Deep brain stimulation (DBS) may be a consideration for treatment of parkinsonism. DBS is high frequency electrical stimulation of very specific and localized areas of the basal ganglia concerned with movement. The two areas of the brain that are FDA approved for stimulation for treatment of parkinsonism and dystonia are the subthalamic nucleus (STN) and the globus pallidus interna (GPi). Leads are generally implanted bilaterally and symmetrically. Each lead contains four electrodes. After placement with stereotaxic surgery, the leads are secured to the skull and the leads burrowed under the skin to the chest or abdomen where the pulse generator (like a cardiac pacemaker) is implanted under the skin. The pulse generator can be interrogated through the skin to determine the battery state as well as the impedence in the electrodes and leads. Also, the pulse generator can be programmed for voltage, frequency, and which of the four contacts in each lead are stimulated. Thus the exact location of the stimulation as well as the parameters of stimulation can be adjusted.

DBS is not for every patient with PD. The principle criterion for DBS is that the symptoms and signs of parkinsonism which are the reason to consider DBS are responsive to levodopa. This levodopa response may be brief and accompanied by dyskinesia but if there is no response to levodopa, DBS is unlikely to be effective. DBS is very likely to help levodopa-sensitive limb bradykinesia, rigidity, and tremor and to decrease dyskinesia. It reduces motor fluctuations in response to levodopa although fluctuations may persist but be less dramatic. Levodopa doses may be decreased but few patients are able to completely stop levodopa. On the other hand, the effects of DBS on balance and gait are less predictable. As reviewed in each chapter, DBS may improve standing postural sway (GPi) and increase gait speed and stride length but does not improve postural reaction and gait stability. Freezing of gait is rarely improved and sometimes induced by DBS.[36] Postural instability before DBS surgery has been linked to poor outcomes after DBS.[37] Operative complications of DBS surgery, such as stroke, hemorrhage, or seizures are uncommon (less than 5%).[38] Other complications of either the lesions produced by lead placement or by the stimulation are effects on balance, cognition, speech, and interest (i.e., produce apathy).[39] These later complications are more common with STN stimulation and are a reason to choose GPi stimulation in many cases.

These considerations require careful evaluation of patients to determine the likelihood of benefit. Evaluation of the patient's balance control when without antiparkinsonian medications overnight (practical Off) and when the antiparkinsonian medications are effective (On) is necessary to determine the levodopa sensitive features of the disease that are likely to respond to DBS. It is important that the patient understands that this test also indicates the extent of improvement they should expect from DBS, and that balance may not improve or may worsen. Unrealistic expectations for the benefits of DBS should be a contraindication for the procedure. The patient and family must also recognize that the disease will continue to progress although the motor benefits derived from DBS will likely persist. The choice of target for DBS is also important. Some centers only perform STN DBS because a greater reduction in levodopa is possible and the antiparkinsonian effects may be marginally greater. However, those benefits are primarily in limb bradykinesia and rigidity and STN DBS is associated with more balance disorders, as well as apathy and speech disorders.[39] In our center, we prefer GPi DBS in PD patients who have any hint of problems with balance, mood, or speech preoperatively.

F. How can freezing of gait be treated?

Milestone 5. Freezing of gait

Freezing of gait (FoG) is arrest of walking most typically occurring during initiation of walking and turning, but during other walking activities as well. The duration of a FoG episode is often brief, less than a few seconds, but occasionally may last a minute or more. When FoG first appears during the course of the disease, the episodes are a couple of seconds in duration and occur infrequently but with time FoG tends to occur many times during the day and limits the patient's mobility around the home and in the community. FoG occurs during the course of PD in a majority of patients[40] but is not pathognomic of PD as FoG is seen in parkinsonism plus syndromes, cortical small vessel disease, and normal pressure hydrocephalus.[41]

One complication of FoG is falls, typically forward on to the knees and outstretched arms. However, patients also fall in other directions because falls during turning, which especially increases FoG, often results in sideways falls. In addition, at the stage that FoG appears, the patients often have other impairments of balance in addition to FoG. Falls occur not only because the body center of mass (CoM) may move beyond the ability to take a step to stop it. People with FoG show more impaired balance across several domains of balance using the Mini-BESTest and instrumented tests.[42]

Determining the relationship of FoG to the medication states is important. If FoG occurs only during OFF times or is more common when Off, although sometimes occurring when On, adjustment of levodopa dosing to reduce Off periods during the day can be helpful. There are reports that continuous enterally administered levodopa producing relatively constant plasma levodopa concentrations may reduce freezing.[43] Although anxiety is prevalent in patients with FoG,[44] it is not clear if treating anxiety will reduce FoG. A recent study suggests that cognitive rehabilitation may reduce FoG since cognitive networks, such as attention and executive function, may be involved.[45]

The physical therapist should address situations that tend to induce FoG and to think of strategies to avoid or ameliorate FoG.[46,47] Likewise tactics to initiate gait without invoking FoG can be tried. For example, the patient should be taught to pause and consciously weight shift to unweight the stepping leg before beginning to walk. Tricks to initiate gait such as shifting weight from side to side and forward or finding a target on the ground to step on or over will help some people initiate walking. Auditory cues such as counting, clapping by a companion, marching music, or a metronome can initiate and sustain walking in some people. Visual cues, such as tape on the floor, a laser pointer, or a companion's foot placed just in front of the patient's foot may also help FoG. The therapist can experiment with tricks and strategies to overcome FoG but realize that no single trick or cue will work consistently and habituation to any cue may occur. Walking aids, particularly walkers, are often very helpful. The tendency for the walker to get in front of the patient when the steps shorten before FoG or stop all together with FoG can cause falls. Reverse brakes in which the patient squeezes the walker brake handles to release rather than to apply the brakes may reduce this tendency. Walkers may also be equipped with a laser projecting a line on the floor to serve as a visual cue. For patients that are falling forward with FoG, knee pads and wrist braces may reduce trauma. If patients are falling in all directions, hip pads and helmets may be a consideration.

G. How should advanced PD incapable of independent ambulation be treated?

Milestone 6. Independent ambulation is no longer possible

With advanced PD, a stage may be reached when the person is no longer safe walking alone, even with a walker. Maintaining safe mobility is still important for quality of life. Physical therapy should consider wheelchairs, power wheelchairs, and scooters. Many people with PD will not be able to control and move a standard wheelchair effectively. With powered devices, it is important to consider whether the patient has the dexterity to operate the device as well as the cognitive wherewithal and judgment to steer the device without running into furniture, walls, and other people. Power wheelchairs and scooters are relatively large and the home environment will factor in the choice of device. Whatever device is judged most appropriate, the chair or scooter must be fit to the patient's habitus and be comfortable. In addition to wheelchairs or scooters for the home, a lightweight, foldable wheelchair will permit the person to continue to participate in community activities such as meals in restaurants, shopping, and movies. Despite the loss of independent ambulation, walking with assistance (a gait belt contributes to the safety of this exercise) is important to maintain strength and flexibility in the hips and legs. Balance training at this stage should include practicing sitting balance with and without support and with and without a secondary cognitive task. The therapist must teach strategies for transfers from wheelchairs to beds, chairs, and toilet. This may require other assistive devices for the home such as bedside toilet, grab bars, and hospital beds. The size and strength of the caregiver(s) are important considerations for choosing transfer techniques. Methods to avoid pressure ulcers should be instituted if appropriate.

Sitting balance may be more difficult on a soft surface or when the feet are not supported if they rely primarily upon somatosensory information for sitting balance orientation. Sitting balance with eyes closed or with limited visual cues may also be more affected if the patient is visually dependent due to poor kinesthesia. Postural transitions to transfer from one support condition to another and in bed should be practiced with or without a caregiver, as appropriate.

Rehabilitation focused on each domain of balance control

At each milestone of balance decline with progression of PD, balance training approach should be specific for the particular types of balance impairments in each patient. However, there is little evidence that general rehabilitation, such as limited to aerobic, strength, and/or home exercises, improve balance control.[16]

There is also no evidence that training one domain of balance control, such as standing balance (Chapter 3), will improve balance control in another domain, such as postural responses (Chapter 4). Few patients will have deficits in only one balance domain, but each patient is unique in the pattern and severity of their balance impairments. Therefore, physical therapists should focus their rehabilitation on the domains most affecting balance and falls in each patient with PD. Based on a comprehensive assessment of balance domains using the Mini-BESTest, therapists can focus on those domains most affected and most likely to be improved with training, both during therapy sessions and combined with longer-term, community exercises. The Chapters 3–8 review the types of exercise approaches therapists can use to address the most common parkinsonian impairments (bradykinesia, rigidity, poor kinesthesis, freezing, lack of automaticity, and executive cognition) affecting each balance domain: postural alignment, limits of stability, postural responses, postural sway in standing, anticipatory postural adjustments, and dynamic balance during gait and postural transitions.

Highlights

- Balance and gait should be addressed from the time of diagnosis of PD because there are balance deficits at the beginning of the disease and the patient already has an elevated risk of falling.
- There is strong evidence that facility-based, directed balance exercises reduce falls. In addition, exercise may have neuroprotective effects for PD.
- Levodopa partially improves some aspects of balance but may also worsen some aspects. Balance disorders are not entirely dopaminergic sensitive indicating that other neurologic systems are important to balance.
- Falls should trigger a complete review of the circumstances associated with the fall, a reexamination of neurological and general physical signs, reassessment of balance with the BESTest, determination of whether the patient was On or Off at the time of fall and scrutiny of the patient's other medications that could contribute to falls.
- Deep Brain Stimulation (DBS) generally does not help balance although it may improve straight ahead walking and bradykinesia of the limbs.
- Freezing of gait responds to various physical therapy tricks and sometimes to improvement in levodopa-induced motor fluctuations.

- Even when unassisted walking is impossible, it is important to maintain the patients' mobility in the home and community. This usually requires wheelchairs and other assistive devices.

References

1. Christiansen C, Moore C, Schenkman M, et al. Factors associated with ambulatory activity in de novo Parkinson disease. *Journal of Neurologic Physical Therapy* 2017;**41**(2):93–100.
2. Lord S, Godfrey A, Galna B, Mhiripiri D, Burn D, Rochester L. Ambulatory activity in incident Parkinson's: more than meets the eye? *Journal of Neurology* 2013;**260**(12): 2964–72.
3. Saaksjarvi K, Knekt P, Mannisto S, et al. Reduced risk of Parkinson's disease associated with lower body mass index and heavy leisure-time physical activity. *European Journal of Epidemiology* 2014;**29**(4):285–92.
4. Xu Q, Park Y, Huang X, et al. Physical activities and future risk of Parkinson disease. *Neurology* 2010;**75**(4):341–8.
5. Ahlskog JE. Does vigorous exercise have a neuroprotective effect in Parkinson disease? *Neurology* 2011;**77**(3):288–94.
6. Kim IY, O'Reilly EJ, Hughes KC, et al. Integration of risk factors for Parkinson disease in 2 large longitudinal cohorts. *Neurology* 2018;**90**(19):e1646–53.
7. Johansson C, Lindstrom B, Forsgren L, Johansson GM. Balance and mobility in patients with newly diagnosed Parkinson's disease - a five-year follow-up of a cohort in northern Sweden. *Disability and Rehabilitation* 2018:1–10.
8. Baltadjieva R, Giladi N, Gruendlinger L, Peretz C, Hausdorff JM. Marked alterations in the gait timing and rhythmicity of patients with de novo Parkinson's disease. *European Journal of Neuroscience* 2006;**24**(6):1815–20.
9. Chou KL, Stacy M, Simuni T, et al. The spectrum of "off" in Parkinson's disease: what have we learned over 40 years? *Parkinsonism and Related Disorders* 2018;**51**:9–16.
10. Lord S, Galna B, Yarnall AJ, et al. Natural history of falls in an incident cohort of Parkinson's disease: early evolution, risk and protective features. *Journal of Neurology* 2017; **264**(11):2268–76.
11. Frazzitta G, Maestri R, Bertotti G, et al. Intensive rehabilitation treatment in early Parkinson's disease: a randomized pilot study with a 2-year follow-up. *Neurorehabilitation and Neural Repair* 2015;**29**(2):123–31.
12. Schenkman M, Moore CG, Kohrt WM, et al. Effect of high-intensity treadmill exercise on motor symptoms in patients with de novo Parkinson disease: a phase 2 randomized clinical trial. *JAMA Neurology* 2018;**75**(2):219–26.
13. Corcos DM, Robichaud JA, David FJ, et al. A two-year randomized controlled trial of progressive resistance exercise for Parkinson's disease. *Movement Disorders: Official Journal of the Movement Disorder Society* 2013;**28**(9):1230–40.
14. Uhrbrand A, Stenager E, Pedersen MS, Dalgas U. Parkinson's disease and intensive exercise therapy–a systematic review and meta-analysis of randomized controlled trials. *Journal of the Neurological Sciences* 2015;**353**(1–2):9–19.
15. Klamroth S, Steib S, Devan S, Pfeifer K. Effects of exercise therapy on postural instability in Parkinson disease: a meta-analysis. *Journal of Neurologic Physical Therapy* 2016;**40**(1):3–14.
16. Shen X, Wong-Yu IS, Mak MK. Effects of exercise on falls, balance, and gait ability in Parkinson's disease: a meta-analysis. *Neurorehabilitation and Neural Repair* 2016;**30**(6):512–27.
17. Yitayeh A, Teshome A. The effectiveness of physiotherapy treatment on balance dysfunction and postural instability in persons with Parkinson's disease: a systematic review and meta-analysis. *BMC Sports Science, Medicine and Rehabilitation* 2016;**8**:17.

18. King LA, Wilhelm J, Chen Y, et al. Effects of group, individual, and home exercise in persons with Parkinson disease: a randomized clinical trial. *Journal of Neurologic Physical Therapy* 2015;**39**(4):204–12.
19. Dos Santos Delabary M, Komeroski IG, Monteiro EP, Costa RR, Haas AN. Effects of dance practice on functional mobility, motor symptoms and quality of life in people with Parkinson's disease: a systematic review with meta-analysis. *Aging Clinical and Experimental Research* 2018;**30**(7):727–35.
20. Li F, Harmer P, Fitzgerald K, et al. Tai chi and postural stability in patients with Parkinson's disease. *New England Journal of Medicine* 2012;**366**(6):511–9.
21. Yang Y, Qiu WQ, Hao YL, Lv ZY, Jiao SJ, Teng JF. The efficacy of traditional Chinese medical exercise for Parkinson's disease: a systematic review and meta-analysis. *PLoS One* 2015;**10**(4):e0122469.
22. Jacobs JV, Horak FB, Van Tran K, Nutt JG. An alternative clinical postural stability test for patients with Parkinson's disease. *Journal of Neurology* 2006;**253**(11):1404–13.
23. Valkovic P, Brozova H, Botzel K, Ruzicka E, Benetin J. Push-and-release test predicts Parkinson fallers and nonfallers better than the pull test: comparison in OFF and ON medication states. *Movement Disorders: Official Journal of the Movement Disorder Society* 2008;**23**(10):1453–7.
24. Beuter A, Hernandez R, Rigal R, Modolo J, Blanchet PJ. Postural sway and effect of levodopa in early Parkinson's disease. *The Canadian Journal of Neurological Sciences* 2008;**35**(1):65–8.
25. Curtze C, Nutt JG, Carlson-Kuhta P, Mancini M, Horak FB. Levodopa is a double-edged sword for balance and gait in people with Parkinson's disease. *Movement Disorders: Official Journal of the Movement Disorder Society* 2015;**30**(10):1361–70.
26. Chung KA, Lobb BM, Murchison CF, Mancini M, Hogarth P, Nutt JG. Assessment of an objective method of dyskinesia measurement in Parkinson's disease. *Movement Disorders Clinical Practice* 2018;**5**(2):160–4.
27. Schrag A, Choudhury M, Kaski D, Gallagher DA. Why do patients with Parkinson's disease fall? A cross-sectional analysis of possible causes of falls. *NPJ Parkinson's Disease* 2015;**1**:15011.
28. Schaafsma JD, Giladi N, Balash Y, Bartels AL, Gurevich T, Hausdorff JM. Gait dynamics in Parkinson's disease: relationship to parkinsonian features, falls and response to levodopa. *Journal of the Neurological Sciences* 2003;**212**(1–2):47–53.
29. Franchignoni F, Horak F, Godi M, Nardone A, Giordano A. Using psychometric techniques to improve the balance evaluation systems test: the mini-BESTest. *Journal of Rehabilitation Medicine* 2010;**42**(4):323–31.
30. Fasano A, Canning CG, Hausdorff JM, Lord S, Rochester L. Falls in Parkinson's disease: a complex and evolving picture. *Movement Disorders: Official Journal of the Movement Disorder Society* 2017;**32**(11):1524–36.
31. Alzahrani H, Venneri A. Cognitive rehabilitation in Parkinson's disease: a systematic review. *Journal of Parkinson's Disease* 2018;**8**(2):233–45.
32. da Silva FC, Iop RDR, de Oliveira LC, et al. Effects of physical exercise programs on cognitive function in Parkinson's disease patients: a systematic review of randomized controlled trials of the last 10 years. *PLoS One* 2018;**13**(2):e0193113.
33. Diaz-Gutierrez MJ, Martinez-Cengotitabengoa M, Saez de Adana E, et al. Relationship between the use of benzodiazepines and falls in older adults: a systematic review. *Maturitas* 2017;**101**:17–22.
34. Laberge S, Crizzle AM. A literature review of psychotropic medications and alcohol as risk factors for falls in community dwelling older adults. *Clinical Drug Investigation* 2019;**39**(2):117–39.

35. Mansfield A, Wong JS, Bryce J, Knorr S, Patterson KK. Does perturbation-based balance training prevent falls? Systematic review and meta-analysis of preliminary randomized controlled trials. *Physical Therapy* 2015;**95**(5):700–9.
36. Collomb-Clerc A, Welter ML. Effects of deep brain stimulation on balance and gait in patients with Parkinson's disease: a systematic neurophysiological review. *Neurophysiologie Clinique = Clinical Neurophysiology* 2015;**45**(4–5):371–88.
37. Abboud H, Genc G, Thompson NR, et al. Predictors of functional and quality of life outcomes following deep brain stimulation surgery in Parkinson's disease patients: disease, patient, and surgical factors. *Parkinson's Disease* 2017;**2017**:5609163.
38. Sorar M, Hanalioglu S, Kocer B, Eser MT, Comoglu SS, Kertmen H. Experience reduces surgical and hardware-related complications of deep brain stimulation surgery: a single-center study of 181 patients operated in six years. *Parkinson's Disease* 2018;**2018**. 3056018.
39. Rossi M, Bruno V, Arena J, Cammarota A, Merello M. Challenges in PD patient management after DBS: a pragmatic review. *Movement Disorders Clinical Practice* 2018;**5**(3): 246–54.
40. Forsaa EB, Larsen JP, Wentzel-Larsen T, Alves G. A 12-year population-based study of freezing of gait in Parkinson's disease. *Parkinsonism and Related Disorders* 2015;**21**(3): 254–8.
41. Giladi N, Kao R, Fahn S. Freezing phenomenon in patients with parkinsonian syndromes. *Movement Disorders: Official Journal of the Movement Disorder Society* 1997; **12**(3):302–5.
42. Bekkers EMJ, Dijkstra BW, Dockx K, Heremans E, Verschueren SMP, Nieuwboer A. Clinical balance scales indicate worse postural control in people with Parkinson's disease who exhibit freezing of gait compared to those who do not: a meta-analysis. *Gait and Posture* 2017;**56**:134–40.
43. Chang FC, Tsui DS, Mahant N, et al. 24 h Levodopa-carbidopa intestinal gel may reduce falls and "unresponsive" freezing of gait in Parkinson's disease. *Parkinsonism ans Related Disorders* 2015;**21**(3):317–20.
44. Ehgoetz Martens KA, Ellard CG, Almeida QJ. Does anxiety cause freezing of gait in Parkinson's disease? *PLoS One* 2014;**9**(9):e106561.
45. Walton CC, Mowszowski L, Gilat M, et al. Cognitive training for freezing of gait in Parkinson's disease: a randomized controlled trial. *NPJ Parkinson's Disease* 2018;**4**:15.
46. Nieuwboer A, Kwakkel G, Rochester L, et al. Cueing training in the home improves gait-related mobility in Parkinson's disease: the RESCUE trial. *Journal of Neurology Neurosurgery and Psychiatry* 2007;**78**(2):134–40.
47. Nonnekes J, Snijders AH, Nutt JG, Deuschl G, Giladi N, Bloem BR. Freezing of gait: a practical approach to management. *The Lancet Neurology* 2015;**14**(7):768–78.

CHAPTER

10

Future perspectives on balance disorders in PD

A. How can wearable technology improve assessment of balance?

Currently, clinical balance measures of people with PD rely primarily on expert-delivered rating scales of PD signs and patient judgment of severity of symptoms. In a clinical setting, the state-of-the-art is to assess balance in PD with clinical rating scales, such as subscores of the Part III of the Unified Parkinson's disease Rating Scale (UPDRS).[1–4] Specifically, balance and gait are measured with the Postural Instability and Gait Disability (PIGD) subscore (including posture, gait, freezing, sit-to-stand, and the pull test) with scores from 0 (normal) to 16 (severe or unable to stand or walk).[1] In addition, balance assessment scales such as the Mini-BESTest,[5] which includes assessment of postural sway, APAs, APRs, and dynamic balance in gait, have been shown to be sensitive to mild PD,[6] FoG,[7] and response to exercise.[8–10]

Unfortunately, these scales are relatively "course" measures of complex motor behavior and all subjective assessments can easily suffer from tester bias, intrarater reliability,[11] or suffer of ceiling effects. An ideal assessment method should provide objective, quantitative measurements that could be easily translated into simple and useful information.

In the last twenty years, developments in microelectronics have led to design and production of a new generation of small, inexpensive and robust sensors, called Inertial Measurement Units (IMUs) that can be used to measure kinematic parameters of the movements of the body segments.[12] IMUs often consist of triaxial accelerometers, triaxial gyroscopes, and magnetometers that can measure leg, arm, and torso motions while people perform clinical balance tasks or go about doing their daily

https://doi.org/10.1016/B978-0-12-813874-8.00010-6

activities. Particularly, accelerometers and gyroscopes have been used to detect and quantify tremor,[13–16] bradykinesia, and hypokinesia[16] in PD patients. Ambulatory monitoring of quality of mobility (gait bouts and turning) have been designed for healthy young subjects, elderly subjects, and pathological cases.[17] Accelerometers, alone, on the wrist or belt have been used as activity monitors to measure the number of steps and sedentary time or to classify the duration of different body postures.[18,19] Also, recently, kinematic sensors have been used to measure joint angles during movement, including the detection and quantification of dyskinesia,[20,21] and ON-OFF states in subjects with PD.[22]

IMUs have also been used, mainly in research environments, to quantitatively assess the balance domains we described in Chapter 3–8 and have the potential to increase the ability to quantitatively and accurately assess these critical aspects of balance in a clinical environment. Wireless, light weight IMUs can quantify balance and gait more quickly and as accurately as traditional motion capture system in a gait laboratory (Fig. 10.1).

FIGURE 10.1 Subject walking while wearing wirelessly synchronized IMUs (Opals by APDM) on her sternum, lumbar spine, and feet to quantify gait and turning.

For example, accelerometers (with appropriate sensitivity)[23] can substitute for traditional force-plate measures to characterize both postural sway during stance (described in Chapter 3) and anticipatory postural adjustments prior to step initiation (described in Chapter 5). Specifically, an inertial sensor placed on the trunk at the L5 level during a standing balance or step initiation task can provide measures of amplitude and frequency of postural sway, or amplitude and duration of the anticipatory postural adjustments phase prior to a step or prior to a one-leg stand task.[24–28] Importantly, such measures were shown to be reliable,[26,28,29] sensitive to early impairments in PD,[27,30] and sensitive to levodopa medication.[31] To characterize automatic postural responses, an instrumented version of the Push & Release test (Chapter 4), with three IMUs on the trunk at L5 level and on the feet, has been proposed and validated with laboratory gold-standard technologies.[32] This test provides clinicians with objective measures of latency of stepping onset, size of first step, number of steps, and time to recover equilibrium.[32]

Dynamic balance during gait and turning can be measured with IMUs. Even one IMU on the belt[33] can estimate gait speed, stride length, and cadence during daily monitoring. Six IMUs on the feet, belt, sternum, and wrists can measure over 150 measures of gait and turning, including stability of the upper body and arm swing, which are particularly affected by PD. Longer walk versions of the Timed-Up and Go have been instrumented[34–37] to characterize postural transitions such as sit-to-stand and turning (Chapter 6–7), in addition to gait characteristics. Longer walks (2-minute or 6-minute) are being used in clinics and research laboratories to accurately measure gait variability, consistency, and fatigue (Chapter 6).

Thus, objective measures of balance using IMUs have the potential to provide clinicians with accurate, stable, and sensitive biomarkers for longitudinal testing of posture and gait. What is needed to make quantitative measures of balance feasible for clinical practice are automatic algorithms for quantifying balance control during prescribed tasks, age-corrected normative values, composite scores, and user-friendly computer interfaces, so the tests can be accomplished quickly and data stored conveniently in electronic medical records.[11] The number of companies investing in this technology to automatically assess balance and gait in pathological populations is increasing, so we expect to find in the market more and more inexpensive, valid, and reliable systems. A potential user should find information about the validity, reliability, and sensitivity of an IMU-based system to measure the particular domains of balance desired for a patient population. The Movement Disorder's Society Task Force on Technology has summarized and identified the challenges and opportunities in the development of technologies with potential for improving the clinical management and the quality of life of individuals with PD.[38] In addition, they outlined a potential framework for the development, accessibility, and long-term adherence of mobile health technologies to enhance care and research objectives related to PD.[39]

B. How will wearable or embedded technology improve mobility in daily life?

Clinical visits may not accurately represent the quality of a patient's mobility and balance control in their home environment.[40] For example, the "white coat effect" often results in better balance and gait when observed during a short test in the clinic than experienced in daily life.[17] Home diaries capture symptom fluctuations better than clinical measures, but diaries are notoriously unreliable and a burden to the patient.[41] Since the use of clinical scales provides only a snapshot of symptom severity during infrequent clinical visits, repeated measurements are useful in revealing the full extent of the subjects' conditions and avoiding bias while measuring the effects of treatment.[42] Therefore, in addition to the need of objective, observer-independent measures of PD balance impairments during prescribed motor tasks, we need objective measures of balance during normal, daily activities. In the future, ambulatory assessment of mobility during daily life activities could provide measures of mobility function outside the clinic, similar to how a heart rate Holter monitor evaluates cardiac function over days and weeks.

The use of body-worn IMUs allows balance measurements outside the clinic or laboratory environments, for example, at home or outdoors. So far, several research studies have used one or more IMUs to characterize balance and gait over multiple days at home, extracting many of the same gait and turning characteristics as can be measured in the laboratory.[33,43–50]

When measuring mobility in the home environment, it is important to differentiate the"quality" of straight-ahead, steady-state walking and turning to predict fallers and distinguish subtle impairments in neurological diseases, compared to the "quantity" or amount of mobility.[33] For instance, greater step-to-step variability was found in retrospective fallers (healthy and subjects with PD), compared to nonfallers, over 3 days of continuous monitoring, although the number of steps per day were similar between the two groups.[48,50] In addition, among the subjects with PD who did not report falls in the previous year, IMU-derived parameters predicted the time to first fall.[50] Generally, the added value of objective measure of mobility over multiple days compared to clinical tests has been recognized a powerful tool in predicting falls in healthy or pathological populations.[46,47,51] In addition to gait quality, quality of turning (for example, turn duration and peak velocity) is also significantly compromised in recurrent fallers, compared to nonfallers, as recently reported[45] and showed specific impairments in people with PD compared to healthy controls as well as in people with FoG history.[52]

Fig. 10.2 shows similar quantity of mobility (number of strides per hour) between a PD and a control group but very different quality of mobility (turning, variability of foot angle at heel strike) over seven days of monitoring in daily life.

A common question when asking people to wear a sensor, even if small and lightweight, is whether people will remember to wear it and charge it and what is the adherence over many days of monitoring. An alternative to wearable sensors to monitor mobility is the use of passive systems that continuously and unobtrusively measure walking activity in a person's home or community center.[53,54] This kind of approach does not require the resident to wear any devices and usually relies on passive, infrared-based, sensing system for continuously assessing walking speed in the home. Such systems have been reported to measure gait speed in a reliable way, and they have been paired with more general platform to predict cognitive and functional ability in independently living older adults.[55] Interestingly, gait speed had been found to be a precursor of cognitive decline in both healthy people[56] and people with PD.[57]

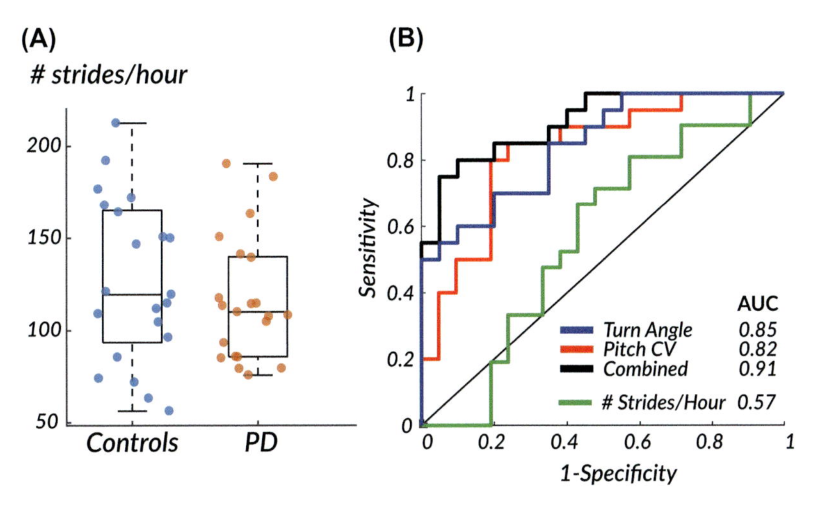

FIGURE 10.2 Despite no difference in number of strides per hour (quantity of mobility) between 20 people with PD and 20 healthy controls over seven days of monitoring with IMUs, quality of mobility, reported as turn angle and variability of pitch angle at heel strike (Pitch CV), can distinguish mobility in people with PD from controls. (A) Box-plot of individual subjects values of average number of strides per hour. (B) ROC (Receiver Operating Characteristic curve) comparing area under the curve (AUC) values among turning angle, variability of pitch angle at heel strike, number of strides/hour in discriminating people with PD from controls. Higher values of the AUC correspond to better discriminatory ability.

Despite promising results in tracking mobility during daily life, many developments and tests are required before a product should be available for clinicians. For example, a consensus on the most important set of gait measures and on a definition of "gait bout" is still missing.[33] In fact gait at home is not as straightforward and stereotyped as in the laboratory. During daily life, the large majority of walking bouts are very brief and often mixed with curved walking and turns. In addition, long-term studies of multiple assessments during daily life over years are still missing, and the reliability of home measures of mobility have not been fully established. For the passive system, the presence of multiple people in the home can still cause confusion, and more importantly only the motor behavior inside the home are recorded, while with wearable technology there is the possibility to measure mobility outside the home as well.

Perhaps the future will involve a merger of these two promising technologies, and/or the wearable technology will be embedded in clothing, allowing greater comfort and usability. In addition, similarly to what we reported for technology to assess balance impairments, the development of user-friendly interfaces and the delivery of easy to read reports will be key for clinical use, as well as use by patients to monitor their own balance and gait.[58,59]

C. How could new technologies improve rehabilitation of balance disorders?

Targeting rehabilitation with objective measures of balance

Technology that provides objective measures of balance domains enables therapists to focus their therapy on the specific domains of balance affected in each individual with PD, whether balance during stance, postural responses, anticipatory postural adjustment, or dynamic balance and turning. Digital technology is especially sensitive for measuring mild balance disorders and for detecting very small changes in balance control with treatment or progression of disease.[60] Postural sway measures while standing under various surface and visual conditions with the Sensory Organization Test of the Equitest have allowed therapists to focus their standing balance therapy on practicing the conditions most difficult for each patient.[17] For example, if standing on a sway-referenced surface results in the most postural instability, the patient would practice standing and walking on unstable surfaces. Since small accelerometers placed on the lower trunk have been shown to be able to replace force plates for measuring postural sway (Chapter 3, and see above), therapists can quickly obtain objective measures of postural sway while standing in the clinic or home or outdoors, under a variety of conditions, such as standing on foam with eyes open or closed.

Body-worn inertial measurement units (IMUs) on the trunk, arms, and feet or ankles can also be used to obtain objective measures of postural responses, anticipatory postural adjustment, or dynamic balance and turning (see above) that therapists can use to focus their therapy. For example, if compensatory stepping responses in response to a Push and Release Test (Chapter 4) are too small, therapists can provide visual cues for larger stepping targets. If APAs are too small, therapists can have patients practice larger, faster lateral weight shifts prior to a step. If walking involves lack of heel strike and small steps, therapists can have patients practice large steps with heel strikes.

Biofeedback, exergaming, virtual reality, and body weight support training

In addition to assessing balance impairments, body-worn sensors can be used to provide real-time biofeedback to patients, or objective feedback to therapists to improve balance control with training. For example, real-time visual or auditory biofeedback about postural sway displacement from an IMU on the belt while standing has been shown to reduce postural sway area.[61] Audio feedback about lateral trunk motion, in which a tilt to the right is represented by a higher tone in the right ear and a tilt to the left is represented by a higher tone in the left ear, has been shown to improve tandem gait.[62] Recently, real-time feedback about gait measures, such as trunk stability, stride length, and arm swing is being tested to see if it improves gait rehabilitation. Feedback from body-worn IMUs can also be used to improve movement quality when patients are exercising on their own. Visual or auditory cues can supplement poor kinesthetic information about body motion. For example, an IMU on the head and sternum could show a patient if they are rotating their head independently of their trunk during exercise. However, it is not clear if improvements seen in standing and walking when using feedback about performance carry over when the biofeedback is unavailable. Longer-term studies of several weeks of practice with feedback from digital technology compared to traditional therapy are needed. Outcome measures for such studies can now also use objective measures from wearable technology in daily life to determine the effectiveness in improving mobility in real-world conditions.

Vibrotactile feedback about body motion can also be used for both training and daily life activities. For example, vibration to the right wrist during the right foot stance phase of gait and to the left wrist during the left foot stance phase of gait has been used to reduce freezing of gait in a laboratory setting ("Vibrogait"[63,64]). Vibration is unobtrusive and more practical for daily life than visual or auditory feedback. However, studies of effectiveness of "Vibrogait" in daily life on measures of freezing taken during daily life are needed.

Researchers are also testing various forms of balance vests or balance belts that provide perceptible cues when the wearer seems liable to fall. For example, Conrad Wall created a device with collaboration from Draper Laboratory in Cambridge, MA, to create a balance system with four tactors (on each side, front and back) orchestrated with a microprocessor that vibrate when the body tilts in that direction (https://spectrum.ieee.org/tech-talk/biomedical/devices/balance-belt.) Another example is an inflating vest with pneumatic actuators that inflate to warn wearers of an impending stumble created by researchers at UCLA's Center for Advanced Surgical and Interventional Technology (CASIT) (https://spectrum.ieee.org/biomedical/devices/vest-helps-keep-balance disorder-patients-from-wobbling/1). Inflating with various pressures and on different sides of the body via a set of 25 balloons around the body give the wearer a physical cue, like a touch that they are listing to port or starboard.

Several studies have utilized gamification of training; technology to assist, resist, or create movements or perturbations; augmented feedback; and virtual reality.[65–73]Many studies have noted improvements in gait and balance following these therapies.[74,75] For example, a large study of 302 older adults, including those with PD showed a reduction of falls after treadmill training plus virtual reality focused on avoiding obstacles. In contrast the randomized group to only treadmill training did not show reduction of falls.[76]

Partial body weight support, either with a harness or with positive pressure pneumatic device over a treadmill that was originally designed to allow patients with spinal cord injury to practice walking is being tested in patients with PD to see if it is superior to treadmill training. Although several studies show improvements in walking distance, UPDRS, gait speed and stride length, most show similar improvements in treadmill training, alone. Short-term gait rehabilitation efficacy of body weight for people with PD with a robotic device (Lokomat-Hocoma Inc., Volketswil, Switzerland) also show improvements in gait and FoG compared to no gait training but not consistently better than conventional, overground gait training without the robotic device.[77]

Although laboratory studies have demonstrated the benefits of practicing responding to a slip or trip to improve automatic postural responses to perturbations (Chapter 4, Postural Responses), new technologies will soon be commercialized to allow physical therapists to safely train postural responses. New platforms, treadmills, and overground robotic harnesses will provide systematic, unpredictable postural perturbations to train postural responses while standing and walking. However, much research is still needed to determine which of these therapies best alleviate which particular domains of balance control, the ideal dose and intensity of these therapies, and long-term retention effects.

Cueing using new, wireless technology, such as immersive reality and virtual reality, show particular promise for improving gait and FoG and is an important area of research. For example, smart phones or smart eye glasses may be used to imbed visual cues onto the floor to overcome FoG events. Music can be played into an ear bud or vibration applied to a limb when a patient's stride length gets too short or freezing is detected from wearable technology.[78–81]

Transcranial direct current to enhance rehabilitation

Recent research has highlighted the potential of transcranial direct current stimulation (tDCS) to complement rehabilitation effects in the elderly and patients with stroke. More recently, the tDCS has been used to enhance neural plasticity for rehabilitation of patients with PD. It has been proposed that tDCS can modulate cortical excitability and enhance neurophysiological mechanisms that compensate for impaired learning in PD. A recent review of the use of TDCS in PD included 10 studies, most of which were sham-controlled. Results confirmed that tDCS applied to the motor or premotor cortex had significant results on UPDRS, time to perform upper limb tasks, and to a lesser extent on cognitive tests.[82]

Four studies found significant effects of tDCS on gait performance.[83–86] Only one study showed the positive effects of tDCS persisted for 4 weeks after intervention.[87] In addition, they showed a significant improvement in the number and duration of freezing episodes during gait, as well as a significant amelioration on the Freezing of Gait Questionnaire. When the effects of bihemispheric tDCS (left and right premotor cortex and M1) with and without contemporary gait training were compared, a significant benefit of combining anodal tDCS with physical training was found and involved moderate to large effect sizes.[88] These relative improvements on gait performance were greater for the combination of stimulation and gait training than the effects of gait training alone. However, 10 sessions of dorsolateral prefrontal cortex (DLPFC) stimulation without gait training failed to show significant effects on walking time.[89] Furthermore, negative results of tDCS alone, without gait training, were found in a study that showed a decrease in walking velocity after stimulation to primary motor cortex (M1) compared to sham.[90] Physical training combined with bihemispheric stimulation to increase excitability in both cortical leg areas seemed to be beneficial for gait.[88] Given that tDCS device is small, relatively inexpensive, portable, and suitable for at-home use, it has the potential to become a usable adjunct to current neurorehabilitation strategies, but not to be used in isolation. However, the physiological mechanism underlying the long-term effects of tDCS on cortical excitability in the PD brain is still unclear, so the ideal brain target is unknown. In addition, the potential effects of tDCS on balance control are less, if ever, studied so improving gait speed may or may not be beneficial to improving functional mobility and reducing falls.

D. Will future medications likely improve balance in PD?

Medical therapy for PD will change over the coming years. What aspects of therapy are likely to change and how will this impact balance issues in people with PD? One long-held goal is to reduce motor fluctuations seen with levodopa by finding better methods to deliver levodopa or alternative medications that can supplant levodopa. Recent introductions to the market of Rytary, an oral extended-release carbidopa/levodopa capsule,[91] and direct duodenal delivery of levodopa by pump[92] are examples of successful methods to reduce fluctuations by reducing the variability of levodopa in the blood. Other techniques are under development including continuous delivery of levodopa subcutaneously and other oral preparations to deliver levodopa continuously.[93]

In addition, gene therapy may improve the response to levodopa by altering the enzyme, aromatic amino acid decarboxylase (AADC), responsible for converting levodopa (a prodrug) into dopamine, in the basal ganglia.[94] Although continuous blood levodopa levels may be achieved, it is likely that there will still be some motor fluctuations because even continuous intravenous delivery of levodopa does not completely abolish fluctuations.[95] The motor fluctuations may be related to variations in entry of levodopa into the brain by the competitive large neutral amino acid transporter that will be affected by the plasma concentrations of other dietary large neutral acids. Alternatively, the non-physiological delivery of levodopa with high brain concentrations of dopamine may preclude a continuous ON response. Thus, although these techniques offer the possibility of reducing motor fluctuations, the fluctuations are unlikely to be abolished. Thus, the effects of ON and OFF fluctuations in balance function will remain the same but with less frequent switches.

Another long-held goal is to develop drugs that would suppress levodopa-induced dyskinesia. At this point in time, the only drug that will partially suppress dyskinesia is amantadine. But there is an active search for other and more effective drugs to directly suppress dyskinesia. The drugs that initially seemed promising have been found to also reduce the effects of levodopa and so have not been brought into clinical practice.[93] Another method to treat levodopa-induced dyskinesia is with deep brain stimulation (DBS) and particularly stimulation of the GPi. This is generally very effective. As dyskinesia is a risk factor for falls, stopping dyskinesia should reduce falls although DBS, itself, may worsen balance as discussed in Chapter 9.

Finally, the ultimate goal in PD therapeutics is to find treatments that will slow or stop the progression of PD. These will likely be drugs to modify pathologic pathways leading to PD or, alternatively, neurotropic growth factors to support the growth, survival, and differentiation of neurons. The pathologic pathways to PD are likely to involve mitochondrial function or the production and handling of alpha synuclein. Trials to modulate the brain concentrations of alpha synuclein are underway and, as there are many mechanisms by which alpha synuclein toxic effects could be modulated, there are many other trials being considered at present.[96] Neurotrophic growth factors have potent effects on the course of parkinsonism in animals, but the trials with direct infusion or of injection with the genes for neurotrophic factors have been disappointing so far, but this is likely related to poor perfusion of the target tissue (the putamen) with adeno-associated viruses (AAV) carrying the neurotrophin gene. These trials are being reconsidered with convection-enhanced perfusion of the putamen with the AAV gene and gadolinium to monitor the extent of perfusion of the putamen.[97]The immediate effects of disease-modifying medications or gene therapies will be to prolong the course of the disease and the need for rehabilitation to maintain mobility over many more years.

E. Could different electrical stimulation targets improve balance in PD?

Deep brain stimulation (DBS) will change in the coming years. With functional MRI studies examining brain circuits for a variety of mental and physical functions, new circuits influencing gait and balance will be discovered, opening up new targets for DBS. New targets identified by fMRI for depression and for cognition dysfunction have already led to pilot trials.[98,99] In the balance realm, small empirical trials have examined DBS in the pedunculopontine nucleus (PPN) in the medullary locomotor region. In PD patients, GABAergic BG output levels are abnormally increased, and gait disturbances are related to abnormal increases in SNr-induced inhibition of the PPN.[100] Given that so much of balance control is nonresponsive to levodopa, the PPN area with its cholinergic outputs may be a good DBS objective for freezing and other balance disorders. Two double-blind small studies showed that a distinct subgroup of parkinsonian patients showed less FoG from PPN DBS. However, there is much controversy at present about its efficacy because the exact best target, stimulation frequency, and ideal patients are unclear.[101] In addition, objective balance outcomes are seldom used to test efficacy. Nevertheless, we anticipate that other targets for DBS will eventually be discovered with efficacy for balance and gait.

In addition to new targets, the delivery of stimulation will change, leading to better control of motor symptoms. This will be by adaptive DBS in which feedback from monitoring electrical patterns from other brain regions or conceivably from measures of balance, gait, tremor, bradykinesia, etc., will feed back to adjust the DBS stimulation parameters in the basal ganglia or thalamus. It is envisioned that adaptive DBS will result in better motor control and also reduce the battery drain that leads to battery replacement within a few years of implantation. Thus, improvement in DBS and new targets may reduce the balance problems associated with parkinsonism.

The spinal cord is a new target for stimulation to improve balance and gait in PD. Small, pilot, open label studies of stimulation over the thoracic, dorsal columns of the spinal cord have recently been shown to improve gait, freezing, and anticipatory postural adjustments in people with severe PD. Well known to reduce neuropathic pain, high frequency stimulation of the thoracic spinal cord is now being tried for people with PD with untreatable FoG and frequent falls. The first experimental report from Fuentes et al.[102] showed that spinal cord stimulation could significantly enhance locomotion in PD models in mice, and more recently in primates, by changing the brain pathological oscillatory activity.[103] So far, less than 10 studies in less than 30 patients suggest that spinal cord stimulation (SCS) may improve treatment-resistant postural and gait disorders in some patients with PD.[104–106] For example, four patients with PD were implanted with stimulators over the upper thoracic spinal cord. SCS improved FoG and size of the APA. However, SCS failed to improve reactive postural responses.[107]

Another potential new target for neural stimulation to improve balance and gait in PD is the vagus nerve.[108] The dorsal motor nucleus of the vagus nerve degenerates in PD, as do cholinergic neurons to which it projects. Noninvasive vagus nerve stimulation (nVNS) in the neck has been used in treatment of epilepsy and migraine and has been shown to activate cholinergic neurons. It has been hypothesized that vagus nerve stimulation could enhance cholinergic activity and therefore, postural sway and gait characteristics that are dopa-resistant and related to fall risk.[108] Clinical trials are needed.

What are the future directions likely to improve balance in PD?

In summary, improved measures of balance impairments in people with PD will lead to: (1) better treatment selection and response monitoring; (2) improved feedback of balance control for patients and physical therapists, and (3) quicker, smaller, less expensive clinical trials to advance new therapies to improve balance (Fig. 10.3). Better balance measures require measures of each domain of balance control (standing posture, postural responses, anticipatory postural control, and dynamic balance during walking) as they likely involve different brain networks. Better balance measures also involve development and testing of objective measures using new technologies. It requires a better understanding of how the brain controls balance and how PD affects this control. It also requires studies of the validity, sensitivity, and responsiveness of new, objective balance measures across the spectrum of disease. Together, these advances in treatment selection and monitoring, rehabilitation training, and new medications will lead to improved balance so people with PD can live longer active lives.

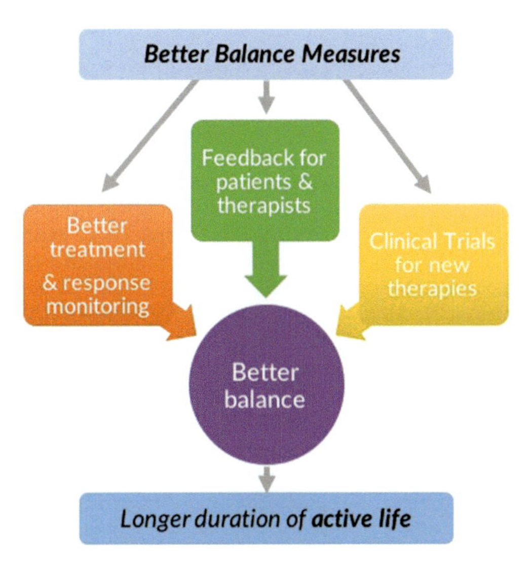

FIGURE 10.3 Schematic of how better balance measures will lead to better balance with long active life with PD.

References

1. Bloem BR, Beckley DJ, van Hilten BJ, Roos RA. Clinimetrics of postural instability in Parkinson's disease. *Journal of Neurology* 1998;**245**(10):669–73.
2. Dibble LE, Lange M. Predicting falls in individuals with Parkinson disease: a reconsideration of clinical balance measures. *Journal of Neurologic Physical Therapy: Journal of Neurologic Physical Therapy* 2006;**30**(2):60–7.
3. Ebersbach G, Baas H, Csoti I, Mungersdorf M, Deuschl G. Scales in Parkinson's disease. *Journal of Neurology* 2006;**253**(Suppl. 4):IV32–5.
4. Haaxma CA, Bloem BR, Borm GF, Horstink MW. Comparison of a timed motor test battery to the unified Parkinson's disease rating scale-III in Parkinson's disease. *Movement Disorders: Official Journal of the Movement Disorder Society* 2008;**23**(12):1707–17.
5. Franchignoni F, Horak F, Godi M, Nardone A, Giordano A. Using psychometric techniques to improve the balance evaluation systems test: the mini-BESTest. *Journal of Rehabilitation Medicine* 2010;**42**(4):323–31.
6. King LA, Priest KC, Salarian A, Pierce D, Horak FB. Comparing the mini-BESTest with the berg balance scale to evaluate balance disorders in Parkinson's disease. *Parkinson's Disease* 2012;**2012**:375419.
7. Bekkers EMJ, Dijkstra BW, Dockx K, Heremans E, Verschueren SMP, Nieuwboer A. Clinical balance scales indicate worse postural control in people with Parkinson's disease who exhibit freezing of gait compared to those who do not: a meta-analysis. *Gait & Posture* 2017;**56**:134–40.
8. Duncan RP, Earhart GM. Randomized controlled trial of community-based dancing to modify disease progression in Parkinson disease. *Neurorehabilitation and Neural Repair* 2012;**26**(2):132–43.
9. Duncan RP, Earhart GM. Are the effects of community-based dance on Parkinson disease severity, balance, and functional mobility reduced with time? A 2-year prospective pilot study. *Journal of Alternative and Complementary Medicine (New York, NY)* 2014;**20**(10):757–63.
10. Rawson KS, Creel P, Templin L, Horin AP, Duncan RP, Earhart GM. Freezing of Gait Boot Camp: feasibility, safety and preliminary efficacy of a community-based group intervention. *Neurodegenerative Disease Management* 2018;**8**(5):307–14.
11. Mancini M, Horak FB. The relevance of clinical balance assessment tools to differentiate balance deficits. *European Journal of Physical and Rehabilitation Medicine* 2010;**46**(2):239–48.
12. Bonato P. Advances in wearable technology and applications in physical medicine and rehabilitation. *Journal of Neuroengineering and Rehabilitation* 2005;**2**(1):2.
13. Frost Jr JD. Triaxial vector accelerometry: a method for quantifying tremor and ataxia. *IEEE Transactions on Bio-Medical Engineering* 1978;**25**(1):17–27.
14. Hoff JI, Wagemans EA, van Hilten BJ. Ambulatory objective assessment of tremor in Parkinson's disease. *Clinical Neuropharmacology* 2001;**24**(5):280–3.
15. Jankovic J, Frost Jr JD. Quantitative assessment of parkinsonian and essential tremor: clinical application of triaxial accelerometry. *Neurology* 1981;**31**(10):1235–40.
16. Salarian A, Russmann H, Wider C, Burkhard PR, Vingerhoets FJ, Aminian K. Quantification of tremor and bradykinesia in Parkinson's disease using a novel ambulatory monitoring system. *IEEE Transactions on Bio-Medical Engineering* 2007;**54**(2):313–22.
17. Horak F, King L, Mancini M. Role of body-worn movement monitor technology for balance and gait rehabilitation. *Physical Therapy* 2015;**95**(3):461–70.

18. Lord S, Rochester L, Baker K, Nieuwboer A. Concurrent validity of accelerometry to measure gait in Parkinsons disease. *Gait & Posture* 2008;**27**(2):357–9.
19. Mactier K, Lord S, Godfrey A, Burn D, Rochester L. The relationship between real world ambulatory activity and falls in incident Parkinson's disease: influence of classification scheme. *Parkinsonism & Related Disorders* 2015;**21**(3):236–42.
20. Burkhard PR, Shale H, Langston JW, Tetrud JW. Quantification of dyskinesia in Parkinson's disease: validation of a novel instrumental method. *Movement Disorders: Official Journal of the Movement Disorder Society* 1999;**14**(5):754–63.
21. Hoff JI, van den Plas AA, Wagemans EA, van Hilten JJ. Accelerometric assessment of levodopa-induced dyskinesias in Parkinson's disease. *Movement Disorders: Official Journal of the Movement Disorder Society* 2001;**16**(1):58–61.
22. Katayama S. Actigraph analysis of diurnal motor fluctuations during dopamine agonist therapy. *European Neurology* 2001;**46**(Suppl. 1):11–7.
23. Chiari L, Dozza M, Cappello A, Horak FB, Macellari V, Giansanti D. Audio-biofeedback for balance improvement: an accelerometry-based system. *IEEE Transactions on Bio-Medical Engineering* 2005;**52**(12):2108–11.
24. Bonora G, Carpinella I, Cattaneo D, Chiari L, Ferrarin M. A new instrumented method for the evaluation of gait initiation and step climbing based on inertial sensors: a pilot application in Parkinson's disease. *Journal of Neuroengineering and Rehabilitation* 2015;**12**: 45.
25. Bonora G, Mancini M, Carpinella I, et al. Investigation of anticipatory postural adjustments during one-leg stance using inertial sensors: evidence from subjects with parkinsonism. *Frontiers in Neurology* 2017;**8**:361.
26. Bonora G, Mancini M, Carpinella I, Chiari L, Horak FB, Ferrarin M. Gait initiation is impaired in subjects with Parkinson's disease in the OFF state: evidence from the analysis of the anticipatory postural adjustments through wearable inertial sensors. *Gait & Posture* 2017;**51**:218–21.
27. Mancini M, Zampieri C, Carlson-Kuhta P, Chiari L, Horak FB. Anticipatory postural adjustments prior to step initiation are hypometric in untreated Parkinson's disease: an accelerometer-based approach. *European Journal of Neurology* 2009;**16**(9):1028–34.
28. Mancini M, Chiari L, Holmstrom L, Salarian A, Horak FB. Validity and reliability of an IMU-based method to detect APAs prior to gait initiation. *Gait & Posture* 2016;**43**: 125–31.
29. Mancini M, Salarian A, Carlson-Kuhta P, et al. ISway: a sensitive, valid and reliable measure of postural control. *Journal of Neuroengineering and Rehabilitation* 2012;**9**:59.
30. Mancini M, Horak FB, Zampieri C, Carlson-Kuhta P, Nutt JG, Chiari L. Trunk accelerometry reveals postural instability in untreated Parkinson's disease. *Parkinsonism & Related Disorders* 2011;**17**(7):557–62.
31. Curtze C, Nutt JG, Carlson-Kuhta P, Mancini M, Horak FB. Levodopa is a double-edged sword for balance and gait in people with Parkinson's disease. *Movement Disorders: Official Journal of the Movement Disorder Society* 2015;**30**(10):1361–70.
32. El-Gohary M, Peterson D, Gera G, Horak FB, Huisinga JM. Validity of the instrumented push and release test to quantify postural responses in persons with multiple sclerosis. *Archives of Physical Medicine and Rehabilitation* 2017;**98**(7):1325–31.
33. Del Din S, Godfrey A, Mazza C, Lord S, Rochester L. Free-living monitoring of Parkinson's disease: lessons from the field. *Movement Disorders: Official Journal of the Movement Disorder Society* 2016;**31**(9):1293–313.
34. Herman T, Weiss A, Brozgol M, Giladi N, Hausdorff JM. Identifying axial and cognitive correlates in patients with Parkinson's disease motor subtype using the instrumented timed up and go. *Experimental Brain Research* 2014;**232**(2):713–21.

35. Weiss A, Herman T, Mirelman A, et al. The transition between turning and sitting in patients with Parkinson's disease: a wearable device detects an unexpected sequence of events. *Gait & Posture* 2019;**67**:224–9.
36. Weiss A, Herman T, Plotnik M, et al. Can an accelerometer enhance the utility of the timed up & go test when evaluating patients with Parkinson's disease? *Medical Engineering & Physics* 2010;**32**(2):119–25.
37. Zampieri C, Salarian A, Carlson-Kuhta P, Aminian K, Nutt JG, Horak FB. The instrumented timed up and go test: potential outcome measure for disease modifying therapies in Parkinson's disease. *Journal of Neurology Neurosurgery and Psychiatry* 2010;**81**(2): 171–6.
38. Espay AJ, Bonato P, Nahab FB, et al. Technology in Parkinson's disease: challenges and opportunities. *Movement Disorders: Official Journal of the Movement Disorder Society* 2016; **31**(9):1272–82.
39. Espay AJ, Hausdorff JM, Sanchez-Ferro A, et al. A roadmap for implementation of patient-centered digital outcome measures in Parkinson's disease obtained using mobile health technologies. *Movement Disorders: Official Journal of the Movement Disorder Society* 2019.
40. Stocchi F, Ruggieri S, Brughitta G, Agnoli A. Problems in daily motor performances in Parkinson's disease: the continuous dopaminergic stimulation. *Journal of Neural Transmission Supplementum* 1986;**22**:209–18.
41. Stone AA, Shiffman S, Schwartz JE, Broderick JE, Hufford MR. Patient compliance with paper and electronic diaries. *Controlled Clinical Trials* 2003;**24**(2):182–99.
42. Isacson D, Bingefors K, Kristiansen IS, Nyholm D. Fluctuating functions related to quality of life in advanced Parkinson disease: effects of duodenal levodopa infusion. *Acta Neurologica Scandinavica* 2008;**118**(6):379–86.
43. El-Gohary M, Pearson S, McNames J, et al. Continuous monitoring of turning in patients with movement disability. *Sensors* 2013;**14**(1):356–69.
44. Mancini M, El-Gohary M, Pearson S, et al. Continuous monitoring of turning in Parkinson's disease: rehabilitation potential. *NeuroRehabilitation* 2015;**37**(1):3–10.
45. Mancini M, Schlueter H, El-Gohary M, et al. Continuous monitoring of turning mobility and its association to falls and cognitive function: a pilot study. *The Journals of Gerontology Series A, Biological Sciences and Medical Sciences* 2016;**71**(8):1102–8.
46. Schwenk M, Hauer K, Zieschang T, Englert S, Mohler J, Najafi B. Sensor-derived physical activity parameters can predict future falls in people with dementia. *Gerontology* 2014;**60**(6):483–92.
47. van Schooten KS, Pijnappels M, Rispens SM, Elders PJ, Lips P, van Dieen JH. Ambulatory fall-risk assessment: amount and quality of daily-life gait predict falls in older adults. *The Journals of Gerontology Series A, Biological Sciences and Medical Sciences* 2015.
48. Weiss A, Brozgol M, Dorfman M, et al. Does the evaluation of gait quality during daily life provide insight into fall risk? A novel approach using 3-day accelerometer recordings. *Neurorehabilitation and Neural Repair* 2013;**27**(8):742–52.
49. Weiss A, Herman T, Giladi N, Hausdorff JM. New evidence for gait abnormalities among Parkinson's disease patients who suffer from freezing of gait: insights using a body-fixed sensor worn for 3 days. *Journal of Neural Transmission* 2015;**122**(3):403–10.
50. Weiss A, Herman T, Giladi N, Hausdorff JM. Objective assessment of fall risk in Parkinson's disease using a body-fixed sensor worn for 3 days. *PLoS One* 2014;**9**(5):e96675.
51. Galperin I, Hillel I, Del Din S, et al. Associations between daily-living physical activity and laboratory-based assessments of motor severity in patients with falls and Parkinson's disease. *Parkinsonism & Related Disorders* 2019.
52. Mancini M, Weiss A, Herman T, Hausdorff JM. Turn around freezing: community-living turning behavior in people with Parkinson's disease. *Frontiers in Neurology* 2018;**9**:18.

53. Hagler S, Austin D, Hayes TL, Kaye J, Pavel M. Unobtrusive and ubiquitous in-home monitoring: a methodology for continuous assessment of gait velocity in elders. *IEEE Transactions on Bio-Medical Engineering* 2010;**57**(4):813–20.
54. Hayes TL, Hagler S, Austin D, Kaye J, Pavel M. Unobtrusive assessment of walking speed in the home using inexpensive PIR sensors. In: *Conference proceedings: annual international conference of the IEEE engineering in medicine and biology society IEEE engineering in medicine and biology society annual conference*, vol. 2009; 2009. p. 7248–51.
55. Kaye J, Mattek N, Dodge H, et al. One walk a year to 1000 within a year: continuous in-home unobtrusive gait assessment of older adults. *Gait & Posture* 2012;**35**(2):197–202.
56. Buracchio T, Dodge HH, Howieson D, Wasserman D, Kaye J. The trajectory of gait speed preceding mild cognitive impairment. *Archives of Neurology* 2010;**67**(8):980–6.
57. Morris R, Lord S, Lawson RA, et al. Gait rather than cognition predicts decline in specific cognitive domains in early Parkinson's disease. *The Journals of Gerontology Series A, Biological Sciences and Medical Sciences* 2017;**72**(12):1656–62.
58. Hansen C, Sanchez-Ferro A, Maetzler W. How mobile health technology and electronic health records will change care of patients with Parkinson's disease. *Journal of Parkinson's Disease* 2018;**8**(s1):S41–5.
59. Klucken J, Krüger R, Schmidt P, Bloem BR. Management of Parkinson's disease 20 years from now: towards digital health pathways. *Journal of Parkinson's Disease* 2018;**8**(s1): S85–s94.
60. Horak FB, Mancini M. Objective biomarkers of balance and gait for Parkinson's disease using body-worn sensors. *Movement Disorders: Official Journal of the Movement Disorder Society* 2013;**28**(11):1544–51.
61. Dozza M, Chiari L, Horak FB. A portable audio-biofeedback system to improve postural control. In: *Conference proceedings: annual international conference of the IEEE engineering in medicine and biology society IEEE engineering in medicine and biology society annual conference*, vol. 7; 2004. p. 4799–802.
62. Dozza M, Wall 3rd C, Peterka RJ, Chiari L, Horak FB. Effects of practicing tandem gait with and without vibrotactile biofeedback in subjects with unilateral vestibular loss. *Journal of Vestibular Research: Equilibrium & Orientation* 2007;**17**(4):195–204.
63. Harrington W, Greenberg A, King E, et al. Alleviating Freezing of Gait using phase-dependent tactile biofeedback. In: *Conference proceedings: annual international conference of the IEEE engineering in medicine and biology society IEEE engineering in medicine and biology society annual conference*, vol. 2016; 2016. p. 5841–4.
64. Mancini M, Smulders K, Harker G, Stuart S, Nutt JG. Assessment of the ability of open- and closed-loop cueing to improve turning and freezing in people with Parkinson's disease. *Scientific Reports* 2018;**8**(1):12773.
65. Barry G, Galna B, Rochester L. The role of exergaming in Parkinson's disease rehabilitation: a systematic review of the evidence. *Journal of Neuroengineering and Rehabilitation* 2014;**11**:33.
66. dos Santos Mendes FA, Pompeu JE, Modenesi Lobo A, et al. Motor learning, retention and transfer after virtual-reality-based training in Parkinson's disease–effect of motor and cognitive demands of games: a longitudinal, controlled clinical study. *Physiotherapy* 2012;**98**(3):217–23.
67. Espay AJ, Baram Y, Dwivedi AK, et al. At-home training with closed-loop augmented-reality cueing device for improving gait in patients with Parkinson disease. *Journal of Rehabilitation Research and Development* 2010;**47**(6):573–81.
68. Gandolfi M, Geroin C, Dimitrova E, et al. Virtual reality telerehabilitation for postural instability in Parkinson's disease: a multicenter, single-blind, randomized, controlled trial. *BioMed Research International* 2017;**2017**:7962826.

69. Gomez-Jordana LI, Stafford J, Peper CLE, Craig CM. Virtual footprints can improve walking performance in people with Parkinson's disease. *Frontiers in Neurology* 2018; **9**:681.
70. Liao YY, Yang YR, Cheng SJ, Wu YR, Fuh JL, Wang RY. Virtual reality-based training to improve obstacle-crossing performance and dynamic balance in patients with Parkinson's disease. *Neurorehabilitation and Neural Repair* 2015;**29**(7):658–67.
71. Mirelman A, Herman T, Nicolai S, et al. Audio-biofeedback training for posture and balance in patients with Parkinson's disease. *Journal of Neuroengineering and Rehabilitation* 2011;**8**:35.
72. Shih MC, Wang RY, Cheng SJ, Yang YR. Effects of a balance-based exergaming intervention using the Kinect sensor on posture stability in individuals with Parkinson's disease: a single-blinded randomized controlled trial. *Journal of Neuroengineering and Rehabilitation* 2016;**13**(1):78.
73. Ginis P, Nieuwboer A, Dorfman M, et al. Feasibility and effects of home-based smartphone-delivered automated feedback training for gait in people with Parkinson's disease: a pilot randomized controlled trial. *Parkinsonism & Related Disorders* 2016;**22**: 28–34.
74. Klamroth S, Steib S, Devan S, Pfeifer K. Effects of exercise therapy on postural instability in Parkinson disease: a meta-analysis. *Journal of Neurologic Physical Therapy* 2016;**40**(1):3–14.
75. Shen X, Mak MK. Repetitive step training with preparatory signals improves stability limits in patients with Parkinson's disease. *Journal of Rehabilitation Medicine* 2012;**44**(11): 944–9.
76. Mirelman A, Rochester L, Maidan I, et al. Addition of a non-immersive virtual reality component to treadmill training to reduce fall risk in older adults (V-TIME): a randomised controlled trial. *The Lancet (London, England)* 2016;**388**(10050):1170–82.
77. Berra E, De Icco R, Avenali M, et al. Body weight support combined with treadmill in the rehabilitation of parkinsonian gait: a review of literature and new data from a controlled study. *Frontiers in Neurology* 2018;**9**:1066.
78. Ford MP, Malone LA, Nyikos I, Yelisetty R, Bickel CS. Gait training with progressive external auditory cueing in persons with Parkinson's disease. *Archives of Physical Medicine and Rehabilitation* 2010;**91**(8):1255–61.
79. Frazzitta G, Maestri R, Uccellini D, Bertotti G, Abelli P. Rehabilitation treatment of gait in patients with Parkinson's disease with freezing: a comparison between two physical therapy protocols using visual and auditory cues with or without treadmill training. *Movement Disorders: Official Journal of the Movement Disorder Society* 2009; **24**(8):1139–43.
80. Olson M, Lockhart TE, Lieberman A. Motor learning deficits in Parkinson's disease (PD) and their effect on training response in gait and balance: a narrative review. *Frontiers in Neurology* 2019;**10**:62.
81. Rochester L, Baker K, Hetherington V, et al. Evidence for motor learning in Parkinson's disease: acquisition, automaticity and retention of cued gait performance after training with external rhythmical cues. *Brain Research* 2010;**1319**:103–11.
82. Broeder S, Nackaerts E, Heremans E, et al. Transcranial direct current stimulation in Parkinson's disease: neurophysiological mechanisms and behavioral effects. *Neuroscience & Biobehavioral Reviews* 2015;**57**:105–17.
83. Benninger DH, Hallett M. Non-invasive brain stimulation for Parkinson's disease: current concepts and outlook 2015. *NeuroRehabilitation* 2015;**37**(1):11–24.
84. Kaski D, Allum JH, Bronstein AM, Dominguez RO. Applying anodal tDCS during tango dancing in a patient with Parkinson's disease. *Neuroscience Letters* 2014;**568**: 39–43.

85. Manenti R, Brambilla M, Rosini S, et al. Time up and go task performance improves after transcranial direct current stimulation in patient affected by Parkinson's disease. *Neuroscience Letters* 2014;**580**:74–7.
86. Manenti R, Cotelli MS, Cobelli C, et al. Transcranial direct current stimulation combined with cognitive training for the treatment of Parkinson disease: a randomized, placebo-controlled study. *Brain Stimulation* 2018;**11**(6):1251–62.
87. Valentino F, Cosentino G, Brighina F, et al. Transcranial direct current stimulation for treatment of freezing of gait: a cross-over study. *Movement Disorders: Official Journal of the Movement Disorder Society* 2014;**29**(8):1064–9.
88. Kaski D, Dominguez RO, Allum JH, Islam AF, Bronstein AM. Combining physical training with transcranial direct current stimulation to improve gait in Parkinson's disease: a pilot randomized controlled study. *Clinical Rehabilitation* 2014;**28**(11):1115–24.
89. Doruk D, Gray Z, Bravo GL, Pascual-Leone A, Fregni F. Effects of tDCS on executive function in Parkinson's disease. *Neuroscience Letters* 2014;**582**:27–31.
90. Verheyden G, Purdey J, Burnett M, Cole J, Ashburn A. Immediate effect of transcranial direct current stimulation on postural stability and functional mobility in Parkinson's disease. *Movement Disorders: Official Journal of the Movement Disorder Society* 2013;**28**(14):2040–1.
91. Greig SL, McKeage K. Carbidopa/levodopa ER capsules (Rytary((R)), numient): a review in Parkinson's disease. *CNS Drugs* 2016;**30**(1):79–90.
92. Olanow CW, Kieburtz K, Odin P, et al. Continuous intrajejunal infusion of levodopa-carbidopa intestinal gel for patients with advanced Parkinson's disease: a randomised, controlled, double-blind, double-dummy study. *The Lancet Neurology* 2014;**13**(2):141–9.
93. Rascol O, Perez-Lloret S, Ferreira JJ. New treatments for levodopa-induced motor complications. *Movement Disorders: Official Journal of the Movement Disorder Society* 2015;**30**(11):1451–60.
94. Christine CW, Bankiewicz KS, Van Laar AD, et al. MRI-guided phase 1 trial of putaminal AADC gene therapy for Parkinson's disease. *Annals of Neurology* 2019.
95. Nutt JG, Carter JH, Lea ES, Woodward WR. Motor fluctuations during continuous levodopa infusions in patients with Parkinson's disease. *Movement Disorders: Official Journal of the Movement Disorder Society* 1997;**12**(3):285–92.
96. Ganguly U, Chakrabarti SS, Kaur U, Mukherjee A, Chakrabarti S. Alpha-synuclein, proteotoxicity and Parkinson's disease: search for neuroprotective therapy. *Current Neuropharmacology* 2018;**16**(7):1086–97.
97. Whone A, Luz M, Boca M, et al. Randomized trial of intermittent intraputamenal glial cell line-derived neurotrophic factor in Parkinson's disease. *Brain: A Journal of Neurology* 2019;**142**(3):512–25.
98. Lv Q, Du A, Wei W, Li Y, Liu G, Wang XP. Deep brain stimulation: a potential treatment for dementia in Alzheimer's disease (AD) and Parkinson's disease dementia (PDD). *Frontiers in Neuroscience* 2018;**12**:360.
99. Riva-Posse P, Choi KS, Holtzheimer PE, et al. A connectomic approach for subcallosal cingulate deep brain stimulation surgery: prospective targeting in treatment-resistant depression. *Molecular Psychiatry* 2018;**23**(4):843–9.
100. French IT, Muthusamy KA. A review of the pedunculopontine nucleus in Parkinson's disease. *Frontiers in Aging Neuroscience* 2018;**10**:99.
101. Hamani C, Lozano AM, Mazzone PA, et al. Pedunculopontine nucleus region deep brain stimulation in Parkinson disease: surgical techniques, side effects, and postoperative imaging. *Stereotactic and Functional Neurosurgery* 2016;**94**(5):307–19.
102. Fuentes R, Petersson P, Siesser WB, Caron MG, Nicolelis MA. Spinal cord stimulation restores locomotion in animal models of Parkinson's disease. *Science (New York, NY)* 2009;**323**(5921):1578–82.

103. Santana MB, Halje P, Simplicio H, et al. Spinal cord stimulation alleviates motor deficits in a primate model of Parkinson disease. *Neuron* 2014;**84**(4):716–22.
104. de Andrade EM, Ghilardi MG, Cury RG, et al. Spinal cord stimulation for Parkinson's disease: a systematic review. *Neurosurgical Review* 2016;**39**(1):27–35 [discussion].
105. Nishioka K, Nakajima M. Beneficial therapeutic effects of spinal cord stimulation in advanced cases of Parkinson's disease with intractable chronic pain: a case series. *Neuromodulation: Journal of the International Neuromodulation Society* 2015;**18**(8):751–3.
106. Yadav AP, Nicolelis MAL. Electrical stimulation of the dorsal columns of the spinal cord for Parkinson's disease. *Movement Disorders: Official Journal of the Movement Disorder Society* 2017;**32**(6):820–32.
107. de Lima-Pardini AC, Coelho DB, Souza CP, et al. Effects of spinal cord stimulation on postural control in Parkinson's disease patients with freezing of gait. *eLife* 2018;**7**.
108. Morris R, Yarnall AJ, Hunter H, Taylor JP, Baker MR, Rochester L. Noninvasive vagus nerve stimulation to target gait impairment in Parkinson's disease. *Mov Disord* 2019 Jun;**34**:918–9. https://doi.org/10.1002/mds.27664. Epub 2019 Mar 19.

Index

'*Note:* Page numbers followed by "t" indicate tables and "f" indicate figures.'

C

D

E

F